MEGALODON UNEARTHED

Unlocking the Secrets behind the Ultimate Prehistoric Shark

EXPLORING THE EVOLUTION AND EXTINCTION *of OTODUS MEGALODON through* FOSSIL EVIDENCE *and* SCIENTIFIC ANALYSIS

Dr. Jay M. Lipoff

Major Scientific Contributions
Dr. Harry M. Maisch, IV

This image was taken in 2018 at a well-known aquarium in the United States. I noticed something amusing besides my two children: the teeth were placed backward in the mouth. I won't specify where it was located, but everyone makes mistakes. Being smart is admitting you were wrong and learning from it.

DEDICATION

First and foremost, I want to thank my family, Julie, Matthew, and Lamden, who have watched me write this book at all hours of the day and night for years. They have always been incredibly understanding and supportive. One day, all my fossils will be yours. Don't forget my Godzillas, too. I love you all from the top of my heart to the bottom of my soul.

I also dedicate this book to all the exciting people I have met over my 35-plus years of diving and fossil hunting.

Courtesy of Hugo Saláis / Metazoa Studio.

ACKNOWLEDGMENTS

My sincerest gratitude and thanks to all of you—my adventurers, artists, authors, fossil hunters, friends, paleontologists, researchers, and all-around geniuses. Your collective efforts and contributions to the research and fossils sections of this book were invaluable.

I am deeply honored to have had the opportunity to work with such a diverse and talented group. Your names are listed here in the order in which I messaged my way to your social media platforms and connected with you. Please be assured that each of you is equally valued, and your efforts have not been taken for granted. I am truly grateful for your involvement in this project.

Thank you to **Michael Konecnik** from Aquanutz Scuba Dive Charters, **Blair Morrow** from Meg Goddess Designs and a Captain at Aquanutz, **Dr. Harry M. Maisch, IV** from the Department of Marine and Earth Sciences, at The Water School Florida Gulf Coast University, **Mark Kostich** from Mark Kostich Photography, **Dr. Kenshu Shimada**—Professor of Paleobiology, from the College of Science and Health at DePaul University **Dr. Bretton W. Kent**, author of Fossil Sharks of the Chesapeake Bay, and from the University of Maryland, **Dr. Victor Perez** from the Environmental Studies Department at St. Mary's College of Maryland, **Dr. Stephen J. Godfrey**—Paleontologist at Calvert Marine Museum, **Dr. Catalina Pimiento** from the Pimiento Research Group, **Jayson Kowinsky** from fossilguy.com, **Tara Gelsomino** from Palmetto Fossil Excursions, **Harrison Miller**—author, **Nilson Clark** from Code Black Fossils, **Russell Brown** from Florida Fossil Hunters, **Danny Case**, **Dr. Albert C. Hine**—Professor Emeritus, College of Marine Sciences, at the University of Southern Florida, **Dr. William B. F. Ryan**—Marine Geologist, **Christina Spence Morgan** from christinamorganillustrations.com, **Jeff Gage**—

photographer for the Florida Museum of Natural History, **Hugo Saláis** from Metazoa Studio, **Kristina Palumbo** from Nautical Necklaces and Aquanutz, **Ryan Meyer**, **Jerome Dinh**, **Nathan Miller**—Science Communicator, Integration and Application Network, UMCES Horn Point Lab, **Dr. Erin Dillon** from the Smithsonian Tropical Research Institute, **Bonnie Farmer** and **Braum Tokarski** from the Calvert Marine Museum, **Dr. Humberto G. Ferrón** from Instituto Cavanilles de Biodiversidad y Biología Evolutiva, University of Valencia, **Skye** and **Joshua Basak** from Palmetto Fossil Excursions, **Cheyenne Hohman** from the University of Kentucky, Kentucky Geological Survey, **Juan Giraldo**—paleoartist, **Frank Mazza** from Paleo Discoveries, **Alex Lundberg**, **Jason Mathias Art Studio**, **Larry Tanenbaum**, **Jeff Lefebvre** from Earth Treasures, **Zachary L. Coverstone**, **Bonnie Nowell**, **Mike Nastasio**, **Dr. Adam Bozeman**, **Garret Hernandez**, **Jaap Roos Art**, **Rick Foresteire**, **Stephen Lee Wenzel** from SWFL Fossil Discovery's, **Martijn Schalk**, **Skylar Vertes**, **Dirk Rouleaux**, **David Ryan**, **Joseph Branin**, **Josh Galloway**, **Ben Beier** and **Marion Becker-Beier**, **Jazzy Jordan** from Jazzy Jordan Art, **Craig Sundell**, **Frank Morrow**, **Michael Tyler Staab** from I Hunt Dead Things, **Mike Jacobsen**, **Maddy Theresa**, **Tee Oghoul**, **Ryan Picou**, **Cooper N. Stephens**, **Matthew Hunt** from Black Water Recovery Divers, **Paul Adams**, **Jason Soward** from Unforgettable Oddities, **Stephen Turner**, **Justin Boorstein** from Every Day I'm Shoveling, **Andrew Ensing**, **Louis Stieffel**, **Mitchell Glenn Winter**, **Clay Cook**, **Jared Shuler** from SCFossils.com, **Dean Rogers**, **Dr. Steven E. Campana** from Life and Environmental Sciences, and University of Iceland and the Canadian Shark Research Lab, **Nathan Foster**, **Sven Reiter**, **Becca LaBrie**, the Charleston Center for Paleontology, **Daniel Reed**, **Melanie Debiais-Thibaud**—Lecturer at Université de Montpellier, **Dr. Paula Dias**, **Gail and Zack Richardson**, **John Taylor** from SharksTeeth.com, **Daryl Serafin**, **Tommy Royal**, **Erin Osborne** from the Charleston Center for Paleontology, and anyone else I might have forgotten.

SPECIAL THANKS

Where do I begin? During one of my trips to Florida to dive with Aquanutz, I met an intriguing fellow named Dr. Harry M. Maisch, IV. Like several other divers, he was on the boat sharing dive stories, but Captains Mike Konecnik and Blair Morrow knew him well. He was intelligent and had a charm that reminded me of a young Sean Connery as James Bond, minus the Scottish accent.

After a few hours of diving, I realized he was a cool guy. Once I discovered how knowledgeable he was, I added him to my list of extraordinary contacts.

I learned more about him from Blair and Captain Mike of Aquanutz, and kept in touch because it is easy to make friends on a dive boat. A bonus is that he's a great resource for bouncing questions off or getting help identifying a fossil, and he's an approachable person. Everyone should have the chance to be on a dive boat with him, and he just found a huge section of a Colombian Mammoth tusk.

When I decided to write a book with a well-researched section on *O. megalodon*, I contacted Harry. I asked him a few questions and explained what I was working on. Without hesitation, he offered to review my information to help keep me on track and ensure that the correct data was provided. The internet pages are filled with so-called "experts." I told him I needed a moment to think about it. YES!

It's akin to writing a book about being a Jedi and having Yoda review your manuscript. Dr. Maisch, is a well-researched and respected paleontologist among his peers and the fossil community. He has co-authored several remarkable studies with other world-renowned paleontologists, enhancing our understanding of the latest theories and intricate details of megalodon life.

One such study was released just before this book was finished. It turned everything the fossil community knew about *O. megalodon* on its head—or is it its tail? So, I went back to the computer for rewrites. It was mildly frustrating, but simultaneously exciting and incredibly fortunate to have access to the latest theories on this incredible giant. Plus, alongside Dr. Kenshu Shimada and 27 other brilliant minds who authored this breakthrough study is my friend, Dr. Harry.

Having him as my sounding board and science fact-checker on this project has been a godsend. I am deeply grateful for his generosity of time and effort, and for even considering overseeing my project. He didn't have to do this, but that is who he is, and I will always be thankful.

He frequently reviewed the research aspects and extensive fossil sections of my book to help me clarify fine details and research findings, compiling a comprehensive narrative about the life of *O. megalodon*. Thank you so much, Harry. What do you want to work on next? Ha ha.

BIG THANKS

Thank you to **Michael Konecnik**, **Blair Morrow**, **Danny Case**, **Ryan Meyer**, **Larry Tanenbaum**, **Josh Galloway**, **Joseph Branin**, **Michael Tyler Staab**, and last but not least, **Dr. Adam Bozeman**, for your help. I sought specific images to enhance the reader's understanding of interesting topics and to provide an opportunity to visualize some wonderful fossils along the journey, and they came through big time.

I want to thank Captain Mike, whom I know from diving with Aquanutz Scuba Diving Charters, for allowing me to visit his home and photograph many of the fossils featured in this book. He has been a great friend, teacher, and supporter of this project.

Blair is an avid diver, the creative force behind Meg Goddess Designs, and an endless ball of energy, enthusiasm, and friendship. We hit it off on my first Aquanutz dive; she is the best. There is no stopping her creativity or her thirst for adventure.

I contacted Danny, Ryan, Larry, Josh, Joseph, Mike, and Adam via Facebook to ask for permission to use several images for the book. Like a salesman cold-calling potential clients, I reached out to them as I had to others. They each spent considerable time browsing their amazing collections and emailing me pictures of very particular fossils I was looking for to help complete the book.

Danny provided an impressive selection of images that enabled me to illustrate the evolution of the teeth from *Otodus obliquus* to *Otodus megalodon*. I was fortunate to share the final image with him before he passed away. He will be sorely missed in the fossil community for his unbridled knowledge, passion, and stunning fossil collection.

Ryan, Larry, and Josh went through their stacks of fossils and presented me with remarkable examples of pathological teeth with various colorations and patterns. Joseph provided marvelous images of mastodon fossils. They contributed some beautiful fossils for this book.

Michael T. Staab of I Hunt Dead Things was prepared to dismantle his collection to get the best photos of his finds, so everything was perfect. I told him it wasn't necessary, but I do appreciate a fellow perfectionist. All their collections are the fossils that museums dream of acquiring.

Adam shared a large sample of his fossils. His talent as a photographer to fully convey the beauty of them as artistic masterpieces is unmatched. I appreciate his work and his willingness to share it with me and you. I found other fossil hunters who have sought out his talents to photograph some of their prized possessions. He's that good. Hopefully, he will publish a book of his work for fossil enthusiasts to enjoy.

Thank you all.

CONTENTS

Shark Fin Above Ocean Water.
© iStock. Credit: DigitalStorm.

A Beautiful Day in Venice, Florida. Courtesy of Blair Morrow.
(Meg Goddess Designs, Aquanutz Scuba Diving Charters).

FOREWORD

The megatoothed shark, Megalodon, has long captivated the minds of adults and children alike. Whether regarded as a curiosity or obsession, the physical remains of giant, serrated teeth exceeding 6" long, deriving from a shark more than double the size of a full-grown, extant great white shark (*Carcharodon carcharias*), are truly awe-inspiring.

Over the last several decades, Megalodon has been the focus of numerous scientific publications, sci-fi books, movies, and music, among other media. A deep dive into these sources reveals how our understanding of Megalodon has evolved in light of new discoveries. Megalodon is now known to have belonged to the *Otodus* lineage and is interpreted to have had a slender, elongated body form, possibly attaining lengths up to ~80 feet.

Like modern sharks, *Otodus megalodon* was cartilaginous and continuously produced and replaced its teeth during its lifetime. A single shark can make thousands, possibly tens of thousands, of teeth during its lifetime. Shark teeth are extremely durable, phosphate-rich remains that can withstand extensive amounts of weathering and erosion. As a result, shark teeth can become concentrated in the marine environment, and they are the most common vertebrate fossils collected globally.

The mystery surrounding Megalodon and the science behind our understanding of this extinct giant shark have captivated Dr. Jay Lipoff. Aside from having a passion for fossils and SCUBA diving, Jay has taken his interests further by writing this guide to share his excitement for Megalodon with others.

Dr. Lipoff is an award-winning chiropractor and owner of Back At Your Best Chiropractic & Physical Therapy. Jay is also the founder of Foundation 4 Heroes, a 501(c)(3) nonprofit inspiring children and honoring Veterans. In his spare time, Jay has also written multiple publications, including *Back At Your Best: Balancing the Demands of Life with the Needs of Your Body*, and three children's books: *Are You Ready To Be A HERO?*, *Super Coco: Will You Be My Friend?*, and *Donny the Megalodon and the JAWsome Miocene Adventure*.

MEGALODON UNEARTHED: Unlocking the Secrets behind the Ultimate Prehistoric Shark presents a comprehensive compilation of information about Megalodon, drawn from fossil evidence and scientific analyses, in a clear and simplified manner. This book also distinguishes between facts and myths, using current data from published, peer-reviewed scientific literature.

Dr. Lipoff requested permission to utilize numerous images published in peer-reviewed journal articles and included many photographs of fossil specimens in this book. The majority of these specimens are from publicly accessible museum collections as well as private collections of amateur paleontologists.

Photographs of fossils readily provide readers with examples of how fossil preservation varies, thereby reinforcing the unique nature of each fossil specimen. Many collectors who have donated unique fossils to museums have played a crucial role in reshaping our understanding of Megalodon and other extinct animals.

With a sharp eye and a little luck, you might make the next big discovery!

—Harry Maisch IV, Ph.D.

Harry Maisch IV, Ph.D., is an Instructor of Marine and Earth Sciences at The Water School at Florida Gulf Coast University. His research primarily focuses on fossil shark, ray, and fish remains from the Late Mesozoic and Cenozoic eras, and he has been collecting marine vertebrate fossils across the USA for over three decades. As an avid SCUBA diver, part of Harry's dissertation involved documenting the fossil shark assemblage from the "meg ledges" in Onslow Bay, NC. He has also found and described microteeth belonging to a new species of stingray from the early Paleocene of Arkansas. Currently, Dr. Maisch's research is centered on the diverse shallow marine fossil assemblages of central and southwest Florida.

Dr. Harry Returning to the Boat with a Nice Meg.
Courtesy of Blair Morrow (Meg Goddess Designs and
Aquanutz Scuba Diving Charters).

Megalodon. Courtesy of Juan Giraldo.

PREFACE

In the early days of dinosaur discoveries, very little was known about these fascinating creatures, and museums were eager to showcase their findings for the public to admire. It drove sales at the box office. This was the Heroic Age of paleontology, a period during which explorers engaged in fierce battles to uncover the most dinosaur specimens.

Unfortunately, these new finds didn't come with instructions, and assembling the fossils was based on their best guesses of a subject they had never seen alive. In some amusing instances, they made incorrect assumptions, such as believing dinosaurs were slow and unintelligent. They speculated about how dinosaurs moved and posed them with their tails dragging behind them on the ground.

Based on partial findings, some dinosaurs were depicted as sprawled out, resembling turtles or lizards, such as Therizinosaurus, before it was understood that the species stood upright. In one case, *Apatosaurus louisae* was displayed at the Carnegie Museum with the head of a Camarasaurus for 45 years until the 1970s.

They opted to use what they had rather than leave a headless dinosaur. To this day, incredible discoveries continue to amaze us, including the revelation that some dinosaurs had feathers and were fast, intelligent, and evolving, as well as nurturing their young.

Likewise, we didn't know much about meteor impacts and ejecta, heat, sound, and wind. That all changed after witnesses of the Tunguska Impact in Russia on June 30, 1908, shared invaluable firsthand accounts of sonic booms, wind bursts, heat, and falling debris.

Tunguska Event. Photo by Leonid Kulik.

This allowed scientists to make better informed assumptions about other impacts, such as the Barringer Crater in Winslow, Arizona, or the Chicxulub crater in the Yucatan Peninsula, which contributed to the extinction of the dinosaurs.

Regarding the disappearance of dinosaurs, a site called Tanis in North Dakota is the latest fossil discovery that will help us understand the immediate aftermath of the Chicxulub impact.

The preservation and diversity of fossilized creatures demonstrate the power of a tsunami wave and how quickly and far it traveled up the Western Interior Seaway. It may have captured the chaotic 10 to 30 minutes immediately after the impact, when land, sea, and even air animals were violently pushed north and trapped in mud while they were still alive.

This allowed their fossilization process to be detailed and unaffected by scavenging or disarticulation. It's as close as paleontologists could get to having a reporter relay the second-by-second details of the catastrophic event. There are decades of work and revelations to be made from this location.

Likewise, until we find a live Megalodon or unearth a complete or compelling new fossil, our understanding of current fossil discoveries relies on speculation and refinement. Without a full skeleton, including the head and tail, predicting a megalodon's size, habits, and lifestyle is impossible and highly speculative.

Comparisons to existing paleontological (extinct) and neontological (living) species possess only limited similarities. Beyond that, they are educated guesses.

However, we now know much more about megalodons than ever before, thanks to paleontologists and regular fossil hunters like you. Countless discoveries are still to be made, and our understanding will continue to grow.

It's an incredibly exciting time for fossil hunters.

***Otodus obliquus* in Matrix**.
Dr. Jay M. Lipoff.

INTRODUCTION

The world looked very different long ago. This introduction summarizes and highlights some awe-inspiring events that have occurred over millions of years, shaping the Earth as we know it today. Our focus will be on Florida because Venice is known as the "Shark Tooth Capital of the World."

The narrative of our planet's geological history unfolds after a pivotal event: the breakup of the supercontinent Pangea, a colossal landmass that was the combination of Laurasia, predominantly the majority of what is now North America, and Gondwanaland, which comprises what is now Africa and South America.[1,2,3]

Around 200 to 175 million years ago (Mya), a monumental split, driven by plate tectonics, set these chunks of continents on a course toward their current positions,[1,2] akin to slicing a birthday cake and distributing the pieces among friends.

Pangea. Dr. Jay M. Lipoff.

This was the age of plate tectonics, or Phase 1 in the creation of Florida,[3] and set the stage for an amazing transformation.

Florida's future site was in a truly unique location as the continents were sorting out their positions. It was nestled between what we would later recognize as Africa and North and South America, a position that would significantly shape its geological history.

The crustal fragments that formed the foundation beneath early Florida were the Suwannee Basin Block and the Florida-Bahama Block.[4] Based on their composition and a trilobite species discovered in these basement rocks that resembled African fossils more than any North American ones, after drilling an oil test well in Madison County, it would be determined later that Florida was originally part of the African plate.[5]

During the Late to Middle Mesozoic (190 to 66 Mya), sea levels were higher, and a large, deep body of water once covered the region below Georgia, known as the Georgia Seaway or Suwannee Channel.[6]

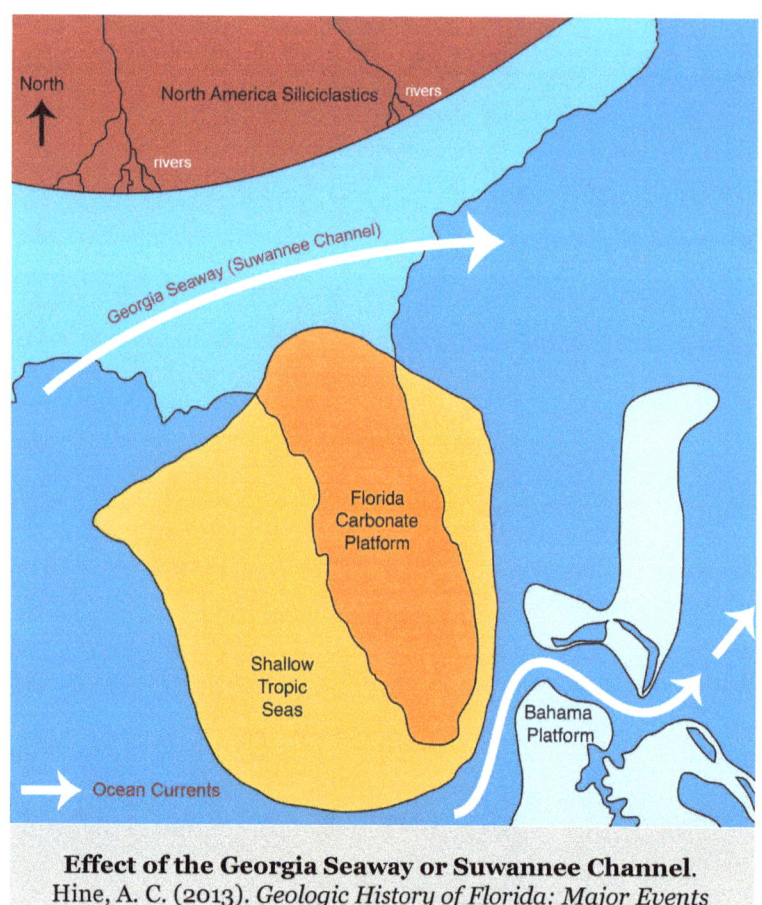

Effect of the Georgia Seaway or Suwannee Channel.
Hine, A. C. (2013). *Geologic History of Florida: Major Events that Formed the Sunshine State*. University Press of Florida. June 18, 2013. Courtesy of Alfred A. Hines. Redrawn by Dr. Jay M. Lipoff.

The Georgia Seaway, a large water channel with a strong current, played a significant role in Florida's geological isolation. This current effectively prevented the siliciclastic sediments, such as quartz sand, silt, and clay, that had been eroding from the Appalachian Mountains for over 30 million years from reaching Florida.[6]

As a result, it isolated the small islands of Florida from the mainland in the North, shaping its unique geological characteristics. This is a testament to the power of natural forces in shaping our world.

From 160 to 23 Mya, the regions of shallow, warm, saline waters provided the perfect conditions for carbonate deposits to accumulate and form the Florida Platform during the Eocene (56 to 33.9 Mya).[7] This was Phase 2, considered the carbonate production age.[4]

As this region flourished with life and death and rising and falling ocean levels, their skeletal remains accumulated along the bottom. Marine animals, including sea urchins, plankton, corals, and shells, collected over time, consolidating into a massive layer of calcium carbonate or a platform.

The reign of the dinosaurs, which spanned millions of years, came to a dramatic end 66 million years ago. A cataclysmic asteroid impact near the Yucatan Peninsula in Mexico contributed to their demise.

Combined with the relentless heat, floods, acid rain, fires, and volcanic activity in India, as well as the climatic changes affecting global sea levels, this series of events drastically altered the environment of the dinosaurs and ultimately led to their extinction.

This impact sent a shockwave across the Florida coast four minutes later,[8] and may have caused sedimentary debris to collapse and mix into a thick wedge of sediments at the escarpment floor.[9]

The Eocene Epoch was like a greenhouse in this region and globally. The air was moist, the water was warm, and the shallow seas above Florida thrived with life. However, by the end of the Eocene, temperatures had fallen, and glaciers began to form, causing sea levels to drop by hundreds of feet.

Forming the Florida Platform. Perez, V. (2022). The chondrichthyan fossil record of the Florida Platform (Eocene–Pleistocene). *Paleobiology, 48(4)*, 1-33. Courtesy of Dr. Victor Perez.

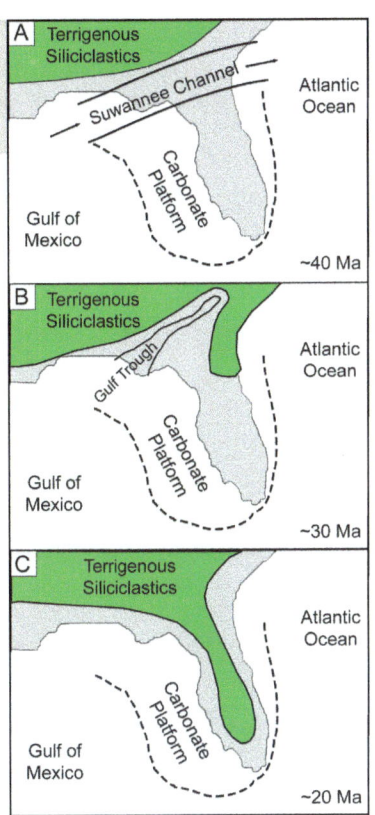

Much like today, our polar regions are melting due to warmer temperatures, and the seas are slowly rising. This process happened frequently during the Cenozoic Era, and multiple major climatic and sea level changes have been documented over the last 24 million years. Near the end of the Eocene, several meteors hit Earth, separated by only a few million years. The Chesapeake Bay bolide meteor, meaning it exploded in the atmosphere just before impact, occurred roughly 35.4 Mya, launching tektites as far away as Georgia and Barbados. It may have sent a tsunami to the Florida Platform 2 hours after it hit.[9]

The Eocene to Oligocene Transition (EOT) is referred to as the greenhouse-to-icehouse transition, during which no glaciers were present. Then, in three transitional phases of cooling, ice sheets grew.[10] As ice volume fluctuates, so do sea levels.

Throughout the late Oligocene (33.9 to 23 Mya), sea levels fell by an estimated 66 feet (20 meters) due to cooler temperatures. Then they dropped an additional 154 to 197 feet (~50 to 60 m) due to the formation of massive amounts of ice.[7]

Florida's Ancient Coastlines

The formation of Orange Island, ~30 mya

Orange Island represents the earliest landmass of what would eventually become Florida. As the global climate cooled during the early to mid-Oligocene epoch (33.9-23 Mya), sea levels fell, exposing the shallow coral reefs and limestone deposits in what is now north-central Florida.

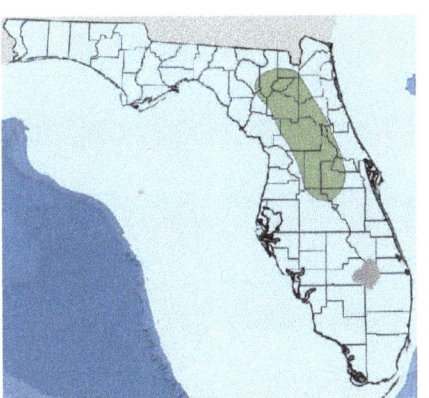

Florida's coastline during the early Miocene, ~23 mya

Over the next several million years, Orange Island continued its rise out of the sea, expanding primarily south and westward. Increasing numbers of terrestrial animals sought refuge in young Florida's expanding grassland savannas and mixed woodland forests while marine life thrived in its shallow, tropical seas.

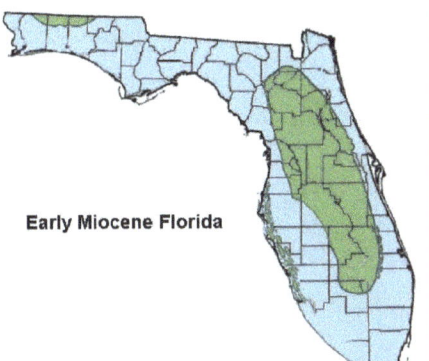

Early Miocene Florida

Florida's coastline at the end of the last glacial maximum, ~18,000 years ago

More commonly known as the "Ice Age," the Pleistocene epoch marked a transition to, on average, significantly cooler global temperatures - a condition climatologists call an Icehouse Earth. These cooler global temperatures led to vast glaciations and the formation of expansive ice sheets, which together covered up to 33% of the planet's landmasses in ice up to 2.5 miles thick at their peak (glacial maxima). This resulted in dramatic drops in global sea levels of up to - 400 feet (~ 125 meters), as was the case a mere ~18,000 years ago when Florida was roughly twice as wide as it is today.

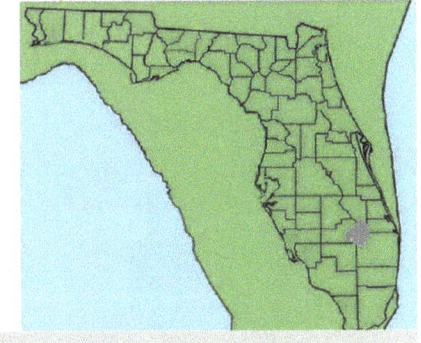

Florida's Ancient Coastlines. From *Geologic History of Florida: Major Events that Formed the Sunshine State*. Courtesy of Albert C. Hine. Modified by Code Black Fossils with descriptions above the images.

As a result, land emerged, small reefs formed, and Orange Island[11] became more visible to the south. Roughly 30 Mya, the once-powerful current of the Suwannee Channel was reduced in size and strength due to falling sea levels and increased sediment production, and it became the Georgia or Gulf Trough.

North America continued to drift north, the Appalachian Mountains rose, and everything changed. Siliciclastic sediments from the north infiltrated the eastern edge of the Florida Platform,[7] from fluvial or river deposition.

This marked the beginning of the end for the Georgia Trough, as sediments accumulated and halted the flow of water, much like a beaver dam. This transformed it from a channel into

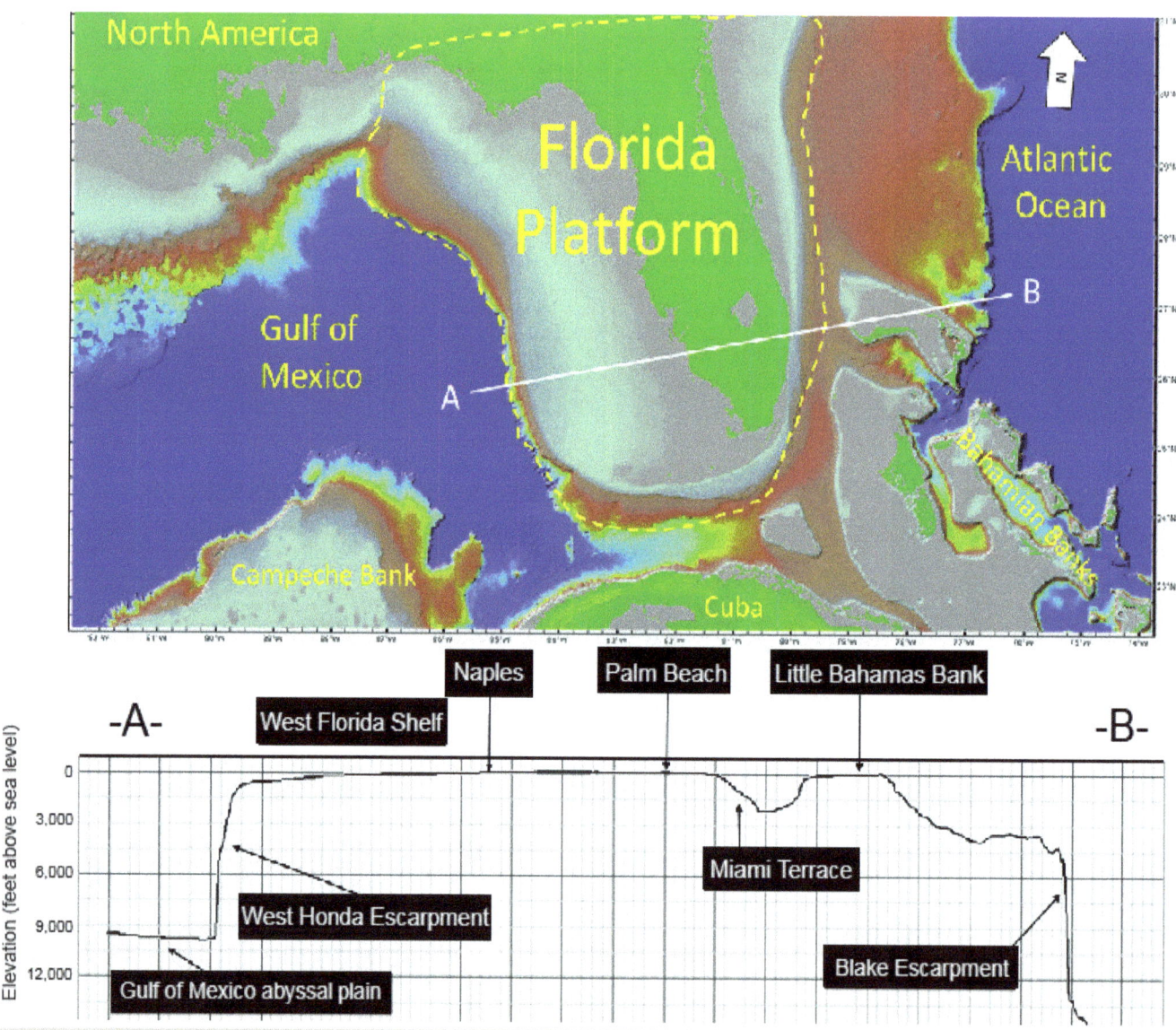

Topography of Florida. Ryan, W. B. F., et al. (2009). Global Multi-Resolution Topography Synthesis. *Geochem. Geophys. Geosyst.*, *10*, Q03014. bit.ly/TopographyofFlorida. Courtesy of William B. F. Ryan.

an inlet. Eventually, the entire waterway was filled, and then rivers ran from the north into Florida, carrying the Appalachian sediments.

While the climate and landscape changed from the Oligocene to the Miocene, so did the animals inhabiting the oceans around Florida. From the Eocene to the Oligocene, the evolution of predatory sharks underwent significant changes.

Sharks with grasping teeth became extinct, and throughout the Miocene, survivors from the orders Lamniformes and Carcharhiniformes developed extremely effective cutting teeth to tear their prey apart.[7]

During the Miocene Epoch, approximately 23 to 5.3 million years ago, the Earth experienced periods of warming and cooling. The tectonic plates of North America moved further north, and the Georgia Trough narrowed as water was recruited to form ice up north.

However, it could no longer deflect the large deposits produced from the continual erosion of the Appalachian Mountains. Now, rivers and streams could transport large quantities of clay, silt, and sand sediments that accumulated and contributed to the enlargement of the Florida Platform's size[12] over 130 million years.[8]

This was called the siliciclastic invasion, which completed Phase 3.[4] The constant sediment deposition continued to creep west and fill the waterway, eventually filling the Georgia Trough approximately 20 Mya.

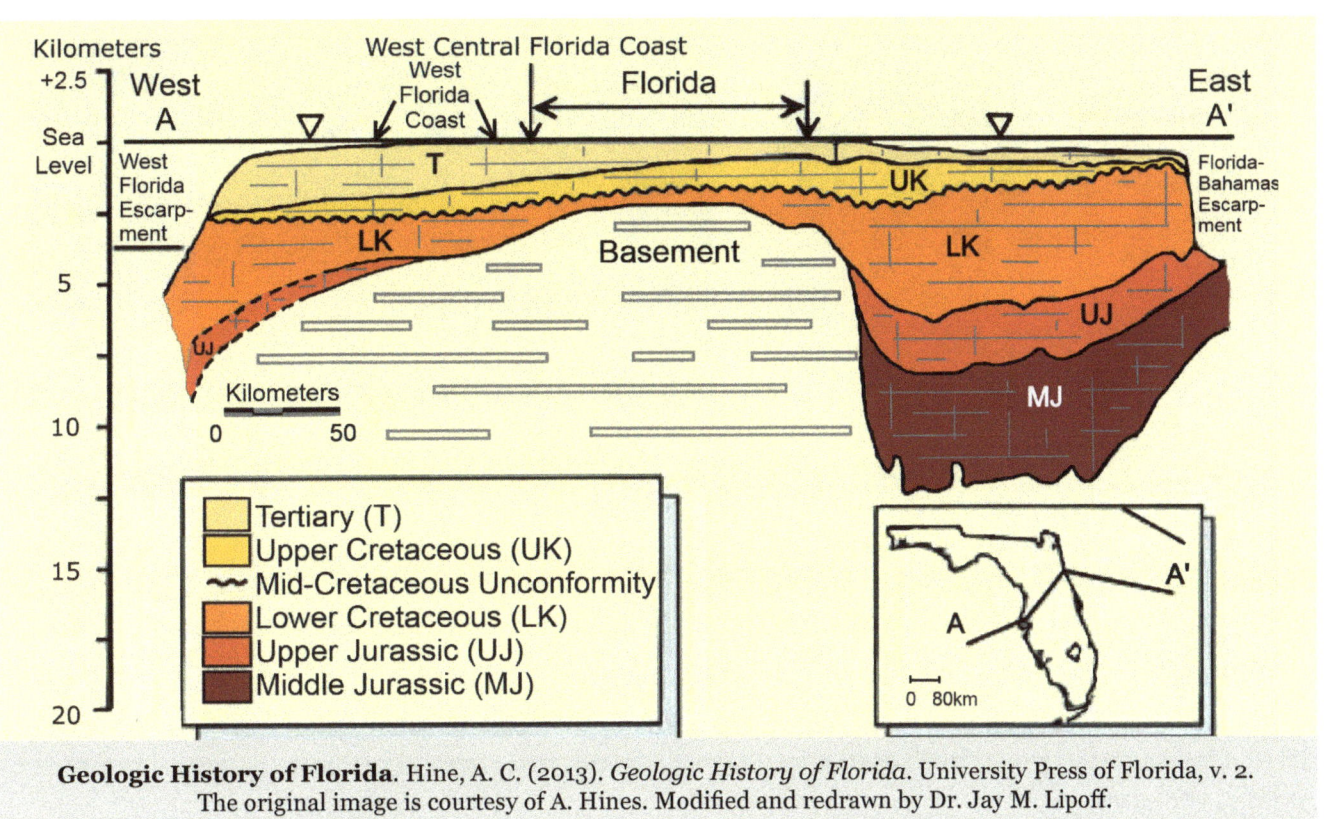

Geologic History of Florida. Hine, A. C. (2013). *Geologic History of Florida*. University Press of Florida, v. 2. The original image is courtesy of A. Hines. Modified and redrawn by Dr. Jay M. Lipoff.

The flat limestone platform, roughly 300 miles (~483 km) wide and 450 miles (~725 km) long, significantly shaped the state of Florida. It would become connected to North America, and the siliciclastic-rich deposits from the mountain range were called the Hawthorn Group.[13]

This influx of deposits blanketed and preserved countless fossils from various species. The shortlist included contributions from camels, saber-toothed cats, dugongs, elephants, ground sloths, horses, mammoths, whales, and teeth from multiple types of sharks.

Rivers, rainwater, and streams would also penetrate the newly exposed limestone, and physical and chemical weathering[7] would create caves, drainage systems, and sinkholes, known as karstification,[14,15] which would form future aquifers.

This substantial drop in sea level during the Miocene led to the emergence of more land and mineral deposits, expanding the mass of Florida to two to three times its current size. Regions like the West Coast extended 100 miles (~161 km) from their present location.

Beyond the shallow shelf to the west is the West Florida Escarpment, a steep 9,000-foot (~2.74 km) drop-off. To the east is a gradual drop of approximately 250 miles (~40 km) to the eventual Blake Escarpment, which descends to depths of up to 16,404 feet (~5,000 m).

In an email with William B. F. Ryan, he mentioned that his team explored the region in the Alvin submersible and that the Blake Escarpment was nearly vertical. It comprises limestones from 160 Mya to 60 Mya, up to the point where the Blake Plateau takes shape.

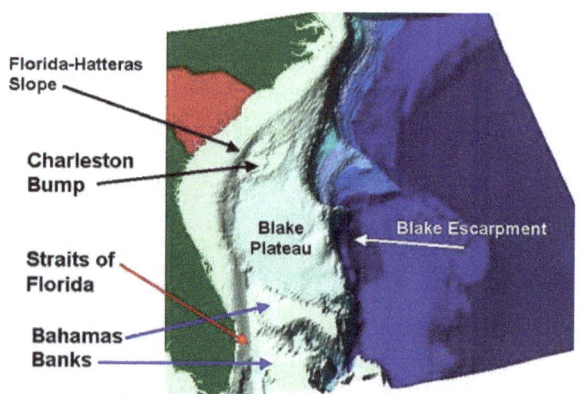

Blake Escarpment. NOAA Ocean Explorer: Islands in the Streams 2001: Digital terrain.

West Florida Escarpment. Courtesy of the NOAA. Office of Ocean Exploration and Research.

Even today, according to statistical data, the average height above sea level in Florida is 100 feet (~30.5 m), and the southern third is only 33 feet (~10 m) or less. It holds the record for the flattest state.

At the end of the Miocene, from 11.6 to 5.3 million years ago, the atmosphere cooled and became drier. Grasslands grew where lush forests once stood, encouraging a wide range of animals to inhabit the area.

Small islands would emerge, forming bridges between landmasses and inviting animals to explore these new regions. Over time, the landscape was inhabited by a diverse array of creatures, including camels, three-toed horses, mastodons, and terror birds, each adapting to the changing environment in its own way.

During this time, the shallow, warm seas covered much of Florida with as much as 100 to 150 feet (~30.5 to 46 meters) of water. This shallow marine habitat likely served as a nursery area for many species and was crucial to the survival of thousands of marine plants and animals that called this place home.

The Laurentide Ice Sheet. Dr. Jay M. Lipoff.

Many adult species raised their young, or the newborns instinctively knew to remain in the safety of the shorelines, because out in the depths, danger lurked in the form of two very large predators: Livyatan, a 60-foot (~18.3-meter) sperm whale, and Megalodon, an equally sized or larger monstrous shark.

At the end of the Miocene and into the Pliocene (~5.3 to 2.58 Mya), ice formation removed ocean water, erasing shallow-water habitats. Thereby, causing the nurseries to disappear.

This led to a modest extinction event, causing nearly 36% of marine megafauna not to survive into the Pleistocene Era.[16] This affirmed that marine life was highly susceptible to global environmental changes.[16]

As the Pleistocene Epoch (~2.58 Mya to 11,700 years ago) began, sea levels rose and fell perpetually due to cooling and the formation of the Laurentide Ice Sheet during the Ice Age. When the glaciers formed, sea levels plummeted by 300 to 400 feet (~91.4 to 122 m).[17]

This brought cooler air, and melting ice flows formed rivers, which carried sediments that modified Florida's coastlines and enlarged ocean terraces. They also continued to preserve many animals beneath these deposits.

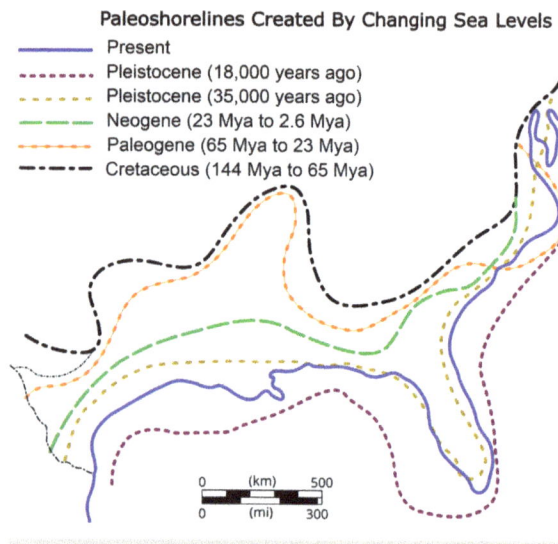

Paleoshorelines Created By Changing Sea Levels

- —— Present
- ······ Pleistocene (18,000 years ago)
- ······ Pleistocene (35,000 years ago)
- – – – Neogene (23 Mya to 2.6 Mya)
- –·–·– Paleogene (65 Mya to 23 Mya)
- –··–··– Cretaceous (144 Mya to 65 Mya)

Paleoshorelines created by changing sea levels. © Wikimedia Commons: Public Domain. NOAA Coastal Services Center. csc.noaa.gov/beachnourishment/html/geo Modified by Dr. Jay M. Lipoff.

These fossils are preserved in the phosphate-rich layers exposed by waterways like the Peace River, making them accessible for discovery and enjoyment.

Here, you may find remains of glyptodonts, dire wolves, terror birds, ground sloths, mastodons, and mammoths, alongside older fossils, including shark teeth.

The phosphate deposits are so incredible and abundant that 80% of the phosphate utilized by the United States and 25% of the world's phosphate needs are mined in Florida.[4]

The entire Atlantic Coast experienced the random rise and fall of the ancient oceans, which is why Maryland, North Carolina, South Carolina, Georgia, and Florida are rich in fossils.

Due to the changing sea-levels, erosion and reworking of the coasts, over 45 taxa or fossil types have been discovered, which were younger the further south they were found. Fossils range from land animals like mammoths and tapirs to sharks and rays.[18] The coasts were alive, and their fossil remains tell the story of their existence.

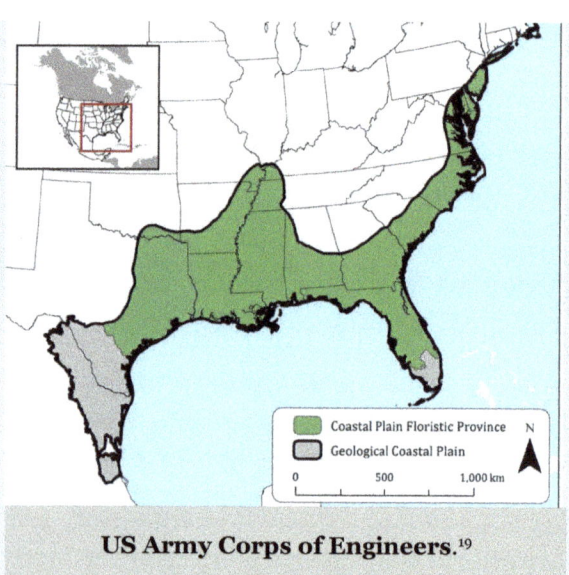

US Army Corps of Engineers.[19]

AS DR. HARRY M. MAISCH, IV BEAUTIFULLY STATED,

"The sea-levels were higher, the shorelines varied, and similar depositional environments/habitats occurred across the Atlantic Coastal Plain. Many shark species also existed in these different basins, but some differences occurred due to bathymetry (depth of water) and water temperature."

For more information on the topic of fossil types found along the Atlantic Coast, I highly recommend this research paper: Maisch, Harry M. IV, et al. (2025). Sharks and Rays (Chondrichthyes: Elasmobranchii) from the Peace River and Tamiami Formations (Late Miocene–Early Pliocene) on the Submerged Continental Shelf near Venice, Florida, USA.

A Friend's Collection on Display.
Dr. Jay M. Lipoff.

CHAPTER ONE

Before Megalodon

Classification of Fish

Approximately 32,000 fish species are in fresh (12,000)[1] and saltwater worldwide.[2] They can be divided into two main classes and vary significantly in identifying characteristics, such as their size, shape, mouth position, scales, gills, and fins.

The first is the jawless fish (Agnatha), like lampreys, which are primitive and lack paired fins.

The second is Gnathostomata, which comprises jawed vertebrates and accounts for over 99% of all living species, including humans.[3,4] They have paired appendages, a skull, and vertebrae;[5] this lineage may go back as far as 420 Mya.[6]

There is an extinct class, the Placodermi, like *Dunkleosteus terrelli*, and two current classes of living jawed fish, classified under Gnathostomata.[6] The first is Osteichthyes, or bony fish, including sturgeon and eels, with two subclasses. One is the ray-finned fish (Actinopterygii), such as tuna, perch, and salmon. Roughly 93% of fish species fall into this group. The other is the lobe-finned fish (Sarcopterygii), such as lungfish and coelacanths.

Labels: Nostril, Eye, Anterior dorsal fin, Posterior dorsal fin, Caudal fin, Buccal tunnel, Head, External gill slits, Trunk, Cloacal aperture, Tail

Lamprey Eel. © Public Domain.

The second branch of living jawed fish is Chondrichthyes, or cartilaginous fish, which includes two subclasses: 1) Elasmobranchii, or sharks, skates, sawfish, and rays, and 2) Holocephali, or fish with a single gill, such as chimeras, rabbitfish, and elephantfish.[7]

Sturgeon (*Acipenser oxyrinchus oxyrinchus*). © U.S. Fish and Wildlife Services.

Chondrichthyans possess heightened sensory abilities. Their mouths are on the underside or ventral surface of their bodies, and, unlike bony fish, they produce only a few offspring at a time.

Some unique features of elasmobranchs include dermal denticles that minimize drag and offer protection from parasites, the famous ampullae of Lorenzini on their snouts, which detect the electrical fields of prey items, and 5 to 7 pairs of gill slits on the sides of their heads.

Currently, there are **OVER 1,200** elasmobranch species, including sharks, skates, rays, and sawfish, known from our modern oceans.[8]

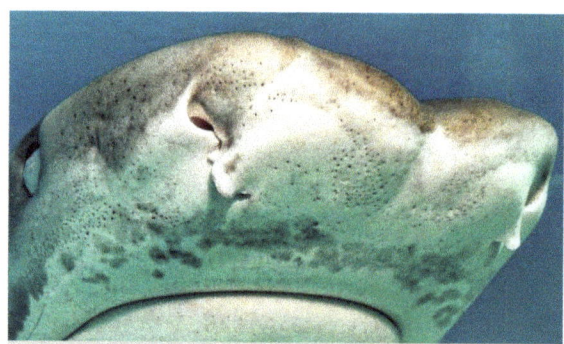

Close-Up of the Ampullae of Lorenzini on the Snout of a Tiger Shark (*Galeocerdo cuvier*).
© iStock. Credit: Serge Melesan.

However, elasmobranch diversity in the recent geologic past may have been even greater. Megalodon falls into this category.

A Brief History of Sharks

Sharks first appeared on Earth 420 million years ago (Mya). The earliest fossilized shark evidence, dating back to the Devonian Period (419.2 to 358.9 Mya), was discovered near Lake Erie in Ohio and in the Stairway Sandstone of Central Australia. However, these fossils do not belong to Megalodon.

Around 252 Mya, at the end of the Permian period, a global extinction known as the "Great Dying" occurred, affecting 90% of all marine species and 70% of land animals.

This event has been linked to massive volcanic eruptions, which have led to increased greenhouse gas emissions, climate change, ocean acidification, and anoxia— the loss of oxygen.

As the planet recovered over the next 10 million years, this presented an opportunity for dinosaurs to evolve approximately 240 to 230 Mya. Numerous sharks persevered through this, too, but not all.

Earth's five mass extinction events and evolution have determined the fate of many species. Sharks like Stethacanthus, Xenacanthus, Scapanorhynchus, and Helicoprion are among the extinct species.

Megalodon is believed to have ancestors dating back to the Cretaceous Period over 100 Mya, although Megalodon itself evolved and thrived between 23 million years ago and 3.6 Mya.[1,2]

Their extinction is determined based on solid fossil evidence of their disappearance. Later dates of 2.6 Mya are based on poor data and fossils derived from older, eroded deposits.[1]

During the Middle Miocene Climatic Optimum (MMCO), approximately 18 to 16 million years ago, temperatures were significantly higher than they are today. Megalodon had evolved into the mesmerizing creature we know and love.

Their numbers increased, dominating the fears of ocean dwellers until they disappeared. How did they survive danger so well, and why did the largest shark[3] in the planet's history,[4] Megalodon, go extinct just a few million years ago?

Xenacanthus from the Triassic Period.
© iStock. Credit: Warpaintcobra.

Late Cretaceous Shark: Hemiscylliidae.
Photo by Chip Clark, Smithsonian, National Museum of Natural History.

Survival Instincts

Research shows that sharks have an innate ability to sense danger. Many sharks left the coast before Hurricane Irma in 2017,[1] like animals fleeing before a tsunami or tornado approaches.

Of the 14 juvenile bull sharks studied, some migrated to different coasts a week earlier, while others left just days to hours before Irma reached shore.[1]

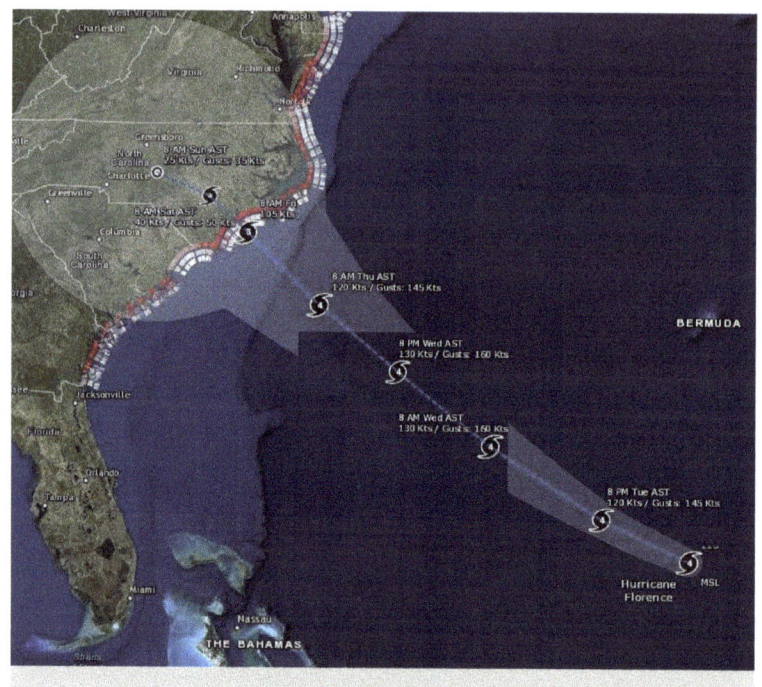

USGS Tracking Hurricane Florence in 2018.
© USGS.

Before Tropical Storm Gabrielle, scientists observed a group of juvenile blacktip sharks migrating to deeper waters. They concluded that a barometric pressure drop alerted the sharks to seek safety. Other fish acted similarly, and all returned to the coastal regions afterward.[2]

These findings align with those of researchers who studied the aftermath of Hurricane Maria's impact on the US Virgin Islands in 2017. Researchers were tracking lemon, nurse, Caribbean reef, and tiger sharks that lived in the Buck Island National Monument Lagoon.

Due to their age, many of these sharks had no prior experience with large storms, and two hours before the storm made landfall, they instinctively sought the safety of deep waters.[3]

However, large tiger sharks did not leave during Hurricane Matthew in 2016, a Category 5 storm. They stayed near the coastal waters of Biscayne Bay and even increased their numbers, likely to scavenge food left after the storm's wrath.[4]

Whether or not Megalodon possessed or utilized the same intuitive behaviors before natural disasters remains to be determined. Some behaviors are impossible to prove without evidence. We can only hypothesize that their behavior may be similar to that of other living sharks.

Sharks have existed longer than humans, so we should respect their intelligence and survival skills. Unfortunately, many are vulnerable to extinction because **human activities result in the annual killing of approximately 100,000,000 sharks**.[5]

To date, over 500 species of sharks are still alive.[5] Given the vastness of the oceans' reach, there may be more that we haven't discovered, and perhaps some that have never been found.

Tropical Storm Edouard in 2002.
Image courtesy of Jacques Descloitres, MODIS
Land Rapid Response Team at NASA GSFC.

The Headington Shark, UK.
© Public Domain. Credit: Steve Daniels.

CHAPTER TWO

Learning from the Past

Fossil Clues

Of course, finding a skeleton would make it much easier to interpret the size of Megalodon. The only problem is that their skeleton rarely fossilizes.

> Unlike a human skeleton, which is hard and designed for weight-bearing, motion, and shock absorption by the discs, a shark's skeleton is composed of pliable cartilage, acts like a spring to store and release energy, and is incredibly efficient at providing flexibility and movement, such as propulsion.

However, there is always a chance, and that improbability happened when 141 preserved vertebrae were miraculously found.[1] This unique specimen is carefully stored at the Royal Belgian Institute of Natural Sciences in Brussels.

There are more parts to a shark's vertebrae, and this will be explained on the next page. The reason most fossil hunters typically only find the centrum or vertebral body is that it is the densest part of the spinal segment.

This shark had 46 growth bands within its vertebrae, similar to rings on a tree limb, indicating that it belonged to a Megalodon with an approximate age of 46.1 years. Researchers used CT imaging to analyze the patterns of these rings further.

It provides a tremendous opportunity and the potential to learn previously unknown facts regarding shark birth size, development, growth rate, and life expectancy, and potentially make inferences about injuries, diseases, and habitat preferences.

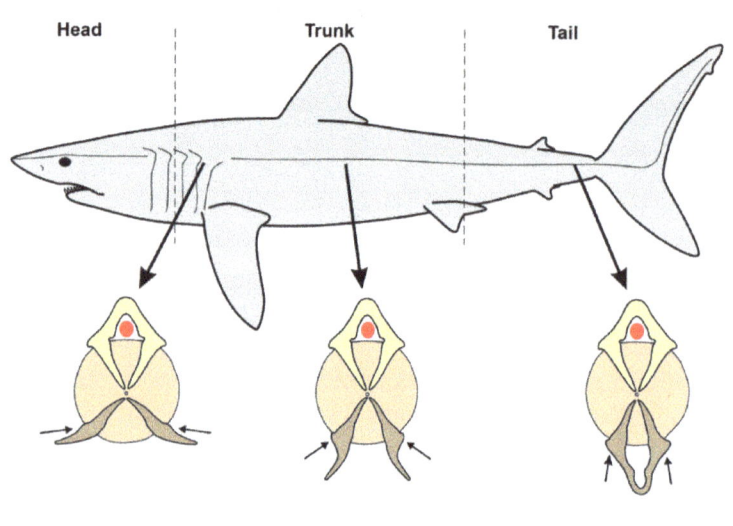

For example, like great white sharks, the life expectancy of the mysterious Megalodon sharks was previously thought to be between 40 and 70 years,[2,3] but may have been as long as 88 to 100 years.[4]

The first image to the left gives a generalized idea of how a shark's vertebrae change shape from the front to the back of the shark.

The next illustration helps identify and understand all the anatomical parts, as well as the reason for the four openings along the sides of the vertebral body.

When the researchers studied the set of 141 fossilized vertebrae, they determined that by counting the rings of the almost 6-inch (~15 cm) diameter fossilized vertebrae,[4,5] the shark would have had a slightly higher growth rate for the first seven years.

After that, the growth rate of this Megalodon was approximately 6.3 inches (~16 cm) per year[1] for the remaining 39 years of its existence.

In Gainesville, Florida, there is an amazing fossil of the ancestor to the modern great white shark. It was discovered in Sacaco, Peru, and consists of 222 teeth, jaw fragments, and 45 vertebrae.[6]

Dr. Gordon Hubbell acquired the specimen and generously donated it to the Florida Museum of Natural History. This fossil was named *Carcharodon hubbelli* in honor of Dr. Hubbell's contributions to science.

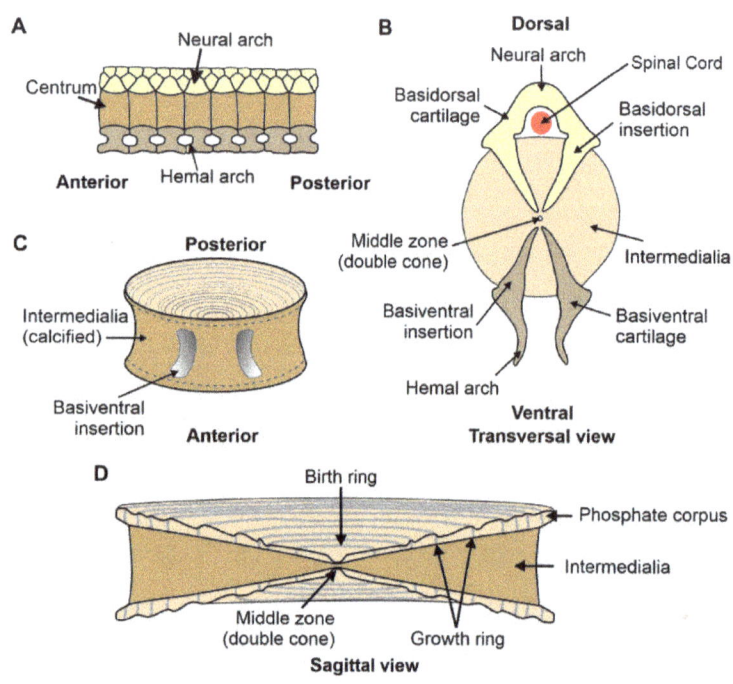

General Anatomy of Shark Vertebrae.
Medeiros, S., Oddone, M. C., Luís, G., Dentzien-Dias, P., & Francischini, H. (2024). Shark (Neoselachii) vertebral centra from the Quaternary of southern Brazil. *Revista Brasileira de Paleontologia. 27(2).* bit.ly/AnatomyofSharkVertebrae. Courtesy of Dr. Paula Dentzien-Dias. Color & Spinal Cord added by Dr. Lipoff.

Dr. Hubbell is not just another fossil hunter who became obsessed with shark teeth after finding them along the beach.

Although a Doctor of Veterinary Medicine, he is best known as an amateur expert in modern and fossil sharks, whom his peers and the public respect.

He also possesses a 7.25-inch (~18.4 cm) tooth from Peru, which is featured later in this book, as well as another 7-inch (~17.8 cm) tooth collected in South Carolina, USA.

Dr. Hubbell is an expert on Megalodon and has an unparalleled collection of shark fossils at his home. He has built his museum, Jaws International, to display his treasures and share his love and understanding with a select group of extremely lucky guests, including fossil experts and enthusiasts.

Growing Up Big

Most sharks give birth to live young, but not all do. Some, like the wobbegong, horn, swell, cat, zebra, bamboo, and Port Jackson sharks, lay eggs.

Carcharodon hubbelli.
Courtesy of Dr. Harry M. Maisch, IV.

46 Associated Megalodon Teeth. From Dr. Hubbell's private collection. Courtesy of Rick Foresteire.

Big Beautiful Shark Vertebra Rings.
Central ring is the size at birth. Dr. Jay M. Lipoff.

Vertebra of *Otodus megalodon*.
Dr. Hubbell's Collection. Courtesy of Louis Stieffel.

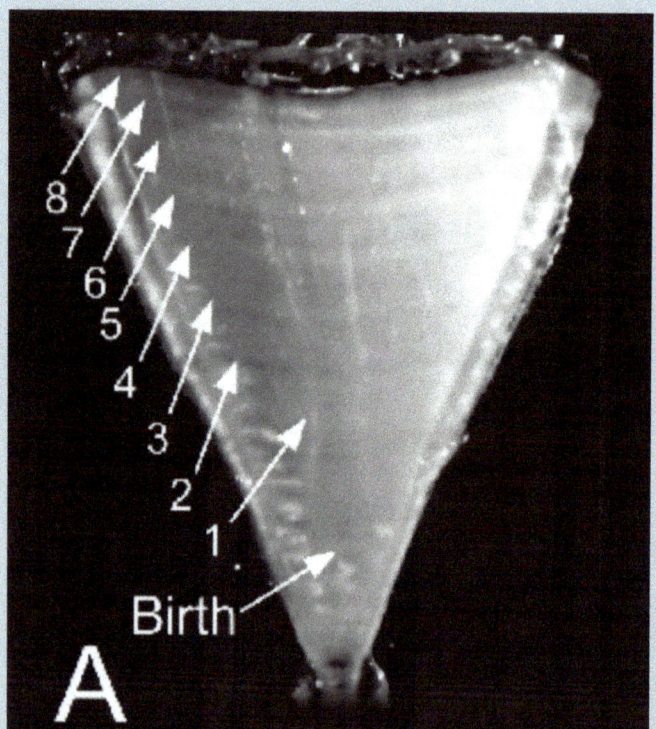

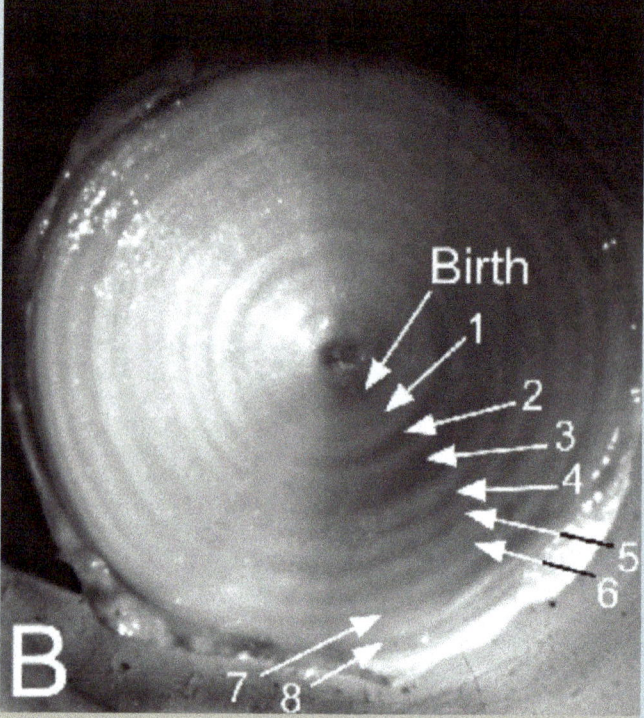

Identifying the Birth and Rings for Each Year of Life of a Shark.
Courtesy of Dr. Steven E. Campana. Life and Environmental Sciences, University of Iceland.
(Canadian Shark Research Lab).

The procreation and perpetuation of the Megalodon species likely had to wait until the female was over 25.[1,2] This delay was necessary because the pup developed inside the mother's womb, a phenomenon known as viviparity,[3,4] and it would result in a sizable delivery.

This seems like a long time; however, the great white shark is believed to reach maturity between 30 to 35 years.[5] Their gestation period, which ranges from 10 to 18 months, is somewhat of a mystery.

According to NOAA Fisheries, a mako shark's gestation period can last 15 to 18 months. The exact gestation period of Megalodon is unknown, but a reasonable estimate would be between 10 and 18 months.

That's a lot of time to plan and pick out an offspring's name, but first, the baby must eat. Therefore, the embryo consumed the nearby unfertilized eggs for nutrition and growth. This type of cannibalism is called oophagy,[4,6,7] and it ensured that the neonate was beautiful and fully capable of surviving.

Ovoviviparity is another survival-of-the-fittest strategy in which eggs hatch and develop inside the mother's body. The embryos begin to eat each other, and at the end of gestation, only a few remain, which are then born.

In the case of Megalodon, there can only be one, and that newborn will be the strongest of the bunch. When this perfect miniature of Mom and Dad was born, and its fins touched water for the first time, it was no vulnerable infant.

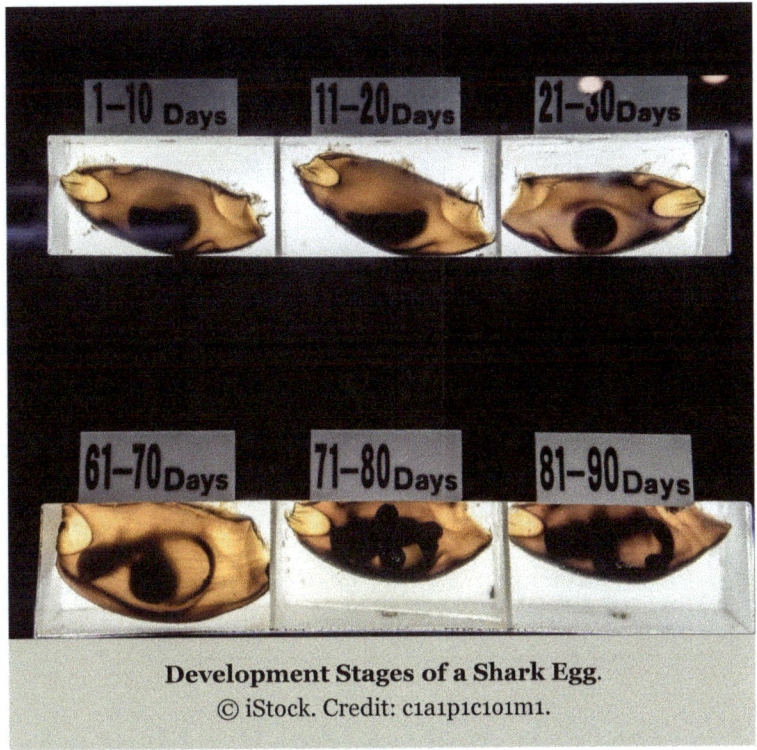

Development Stages of a Shark Egg.
© iStock. Credit: c1a1p1c1o1m1.

Young Shark Swimming Near Shore.
© iStock. Credit: FamVeld.

Previous estimates of the size of a newborn Megalodon ranged from 6 to 7 feet (~1.83 to 2.13 m) long.[1,8] In this model, it was ready to take on the world, but like many young marine species, it probably stayed in shallow, warmer waters for growth, ample food, and safety. Unlike whales, dolphins, and orcas, which stay with their young to nurture and protect them, there were no Megalodon parents to teach them how to survive. The moment it touched water for the first time, it was on its own.

Interestingly, a great white shark mother may give birth every 2 to 3 years to a litter of between 2 and 17 pups, each measuring 4 to 5 feet (~1.2 to 1.5 m) in length. During the first 5 years of life, they grow approximately 12 inches (~0.3 m) per year and are fully capable of hunting and surviving.

At the same time, a study of the Miocene era on the South Coast of Indonesia found the habitat of Megalodon juveniles only ventured to a depth of 0 to 130 feet (~0 to 40 m), and adults went to depths of 263 to 525 feet (~80 to 160 m) deep.[9]

Based on previous findings and our understanding of Megalodons, a research study inferred that an adult Megalodon could weigh over 50 tons or 110,200 pounds (~49,986 kilograms).[10,11] That's close to the weight of 9 elephants.

The latest study suggests that Megalodons may have **grown even larger**, reaching **lengths of 80 feet** (~24.3 m) and **weighing around 92.5 tons** or 207,000 pounds (~94,000 kg), which increased their caloric needs.[12]

The jaw of a large Megalodon could have reached 9 by 11 feet (~2.74 by 3.35 m) when it was wide open and ready to tear into a baleen whale. It may have had a bite force of around 40,000 pounds per square inch (psi).[13]

It is believed that a large *Tyrannosaurus rex* may have averaged around 12,000 psi, a Nile crocodile about 5,000 psi, a great white shark 4,000 psi; a hippopotamus is around 1,800 psi; a bull shark 1,350 psi; and a human only has a bite force between 100 and 150 psi.[14,15]

One study extrapolated measurements from other similar shark species. It concluded that a Megalodon, approximately three times longer than a great white[11] and 15 feet (~4.6 m) bigger than a school bus at 52.5 feet (~16 m), would have the following estimated bodily dimensions.

A head roughly 15.3 feet (~4.65 m) long, a dorsal fin around 5.3 feet (~1.62 m) high, and a tail over twice that length at 12.6 feet (~3.85 m) long.[11] However, the measurements of the elongated dorsal and pectoral fins aligned more with a quick predator that swam for long periods.[16]

A 26-foot (~8 m) juvenile would possess a 7.6 ft. (~2.32 m) long head, 2.7 ft. (~0.81 m) tall dorsal fin, and a 6.3 ft. (~1.92 m) high tail. The newborn of only 9.8 ft. (~3 m) long would be much cuter with a 2.9 ft. (~0.87 m) long head, 1.0 ft. (~0.31 m) tall dorsal fin, and 2.4 ft. (~0.72 m) high tail.[11] It was a mini version but just as deadly, with a low predation risk and an advantage over its competitors.[12]

The sizes of these sharks during their developmental stages vary between studies and scientists. They have also been calculated as adults, being 34.5 feet and up (~10.5 m and up), juveniles having a range between 13 and 33 feet (~4 to 10 m), and neonates being up to 13 feet (~4 m) long.[12]

However, it must be reiterated that many previous conclusions were based on factors such as jaw or tooth width,[16] which naturally vary, and on the assumption that great white and Megalodon sharks had a similar appearance. Some calculations were even made by substituting knowledge of great white shark vertebrae to help reconstruct a complete fossilized megalodon vertebral column. These assumptions may be incorrect.

A recent study disputes this and significantly alters our understanding of Megalodons, presenting numerous new insights into their existence. For instance, the cross-sectional measurement of the Megalodon vertebrae in Denmark was roughly 9 inches (~23 cm) in diameter,[12] and the middle or thoracic spine was also slightly arched, similar to the kyphotic curve found in humans, but less pronounced.

When they counted the rings, they realized that this shark may have entered the world as large as 11.8 to 12.8 feet (~3.6 to 3.9 m) long and had grown at an average rate of 14.72 inches per year (~37.4 cm/yr) during its first seven years.[12]

This growth rate outpaced that of all other predators, giving it a sizable advantage.[12] The remaining life of this animal exhibited a slower growth rate, averaging 10.4 inches (~26.5 cm/yr).[12]

Now, the question is, if the "baby" shark was almost 13 feet long (~3.96 m) at birth,[12] did it need the safety of a shallow, 100 to 300 foot (~30.5 to 91.4 m) deep coastal plain region, a warm environment to survive, or was it comfortable in open waters right from the start?

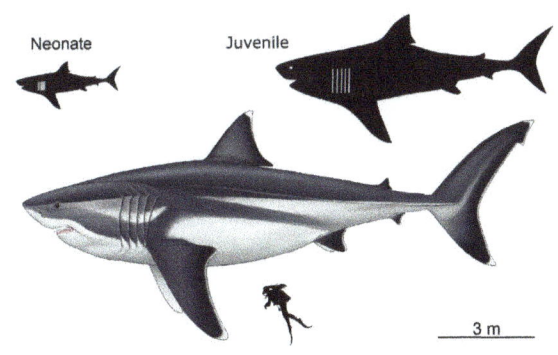

Size Comparison of Megalodon During its Growth. Modified from an illustration by Oliver E. Demuth by Dr. Jay M. Lipoff. Source: bit.ly/MegalodonGrowth.

Baby whales, orcas, and dolphins stay in the deep sea with their mothers. In contrast, sharks do not nurture their young in this manner, and juveniles typically venture off independently, with some potentially thriving in coastal regions.

It seems that Megalodon may have visited the coastal regions, and who could blame them? During the Late Miocene, the water temperature in Florida was approximately 82 ± 2 °F (~27.8 ± 1.1 °C).[17,18] Coincidentally, later, the warmth of the Gulf provided temporary protection against the extreme cooling of the northern environments.

Comparison of Vertebral Diameter, Height, and Species Length

Animal	Average Vertebrae Diameter	Average Vertebrae Height	Animal Length
Megalodon	6 to 9 inches (~15.24 to 23 cm)	6 to 9 inches (~15.24 to 23 cm)	Up to 80 feet (~24.3 m)
Blue Whale	12 to 16 inches (~30 to 40 cm), MAX 3 ft (~0.9 m)	12 to 16 inches (~30 to 40 cm)	98 to 108 feet (~30 to 33 m)
Sperm Whale	12 to 20 inches (~30 to 50 cm)	12 to 20 inches (~30 to 50 cm)	50 to 66 feet (~15.24 to 20 m)
Great White	1 to 2 inches (~2.5 to 5 cm)	0.8 to 2 inches (~2 to 5 cm)	20 feet (~6 m)
Mako Shark	0.4 to 1 inch (~1 to 2.5 cm)	0.4 to 1 inch (~1 to 2.5 cm)	13 feet (~4 m)
Basking Shark	0.4 to 1.2 inches (~1 to 3 cm)	0.8 to 1.6 inches (~2 to 4 cm)	26 to 40.3 feet (~7.9 to 12.3 m)
Human	1 to 2 inches (~2.5 to 5 cm)	0.8 to 1.2 inches (~2 to 3 cm)	70 inches (~1.78 m)

Comparison of Vertebral Diameter, Height, and Species Length. This was created to visualize how megalodon's vertebrae compare to those of other species. © 2025 Dr. Jay M. Lipoff.

They were also well-adapted and thrived in the open waters. Several whale vertebrae have been found along the eastern coastline of the USA with clear predation marks made by Megalodons.[19,20,21]

Megalodon teeth can also be found on the coast. Are they from resident Megalodons or from carcasses that sank or drifted to the floor of the coastal plains, or were they carried by undercurrents that moved the fossils?

Later, we will discuss a large Megalodon found in Maryland that left behind over 150 teeth, some of which were 5 inches (~12.7 cm) in length, to tell its tale.

Full-Grown Megalodon - 80 Feet (~24.38 m)

Great White - 20 Feet (~6.1 m)

Megalodon Neonate - 13 Feet (~3.96 m)

Human - 6 Feet (~1.83 m)

Scaled Images Modified from Shimada et al. (2025).
The exact size and shape of the fins are unknown. © 2025 Dr. Jay M. Lipoff.

Behind Every Great Tooth is a Great Story. What's Yours?
Courtesy of Blair Morrow (Meg Goddess Designs, Aquanutz Scuba Diving Charters).

CHAPTER THREE

Coastal Nurseries

Even though most of us are searching for fossil teeth, don't disregard the smaller ones. They, too, have a story to tell and have been recovered at various sites of proposed nurseries.[1]

The question is whether nurseries existed as safe havens for juveniles, thereby ensuring their survival, or whether warmer waters generally resulted in smaller Megalodon individuals.[1] Perhaps it was a little of both. Smaller sharks in warmer climates with nurseries.

Researchers have discovered prehistoric evidence of a great white shark nursery from the Pliocene Epoch (~4 Mya). After examining their data, including body size distribution interpreted from fossil teeth from three different South American regions, these researchers concluded that juvenile sharks dominated a location in Coquimbo, Chile. In contrast, subadult and adult sharks proliferated more near Caldera, Chile, and Pisco, Peru.[2]

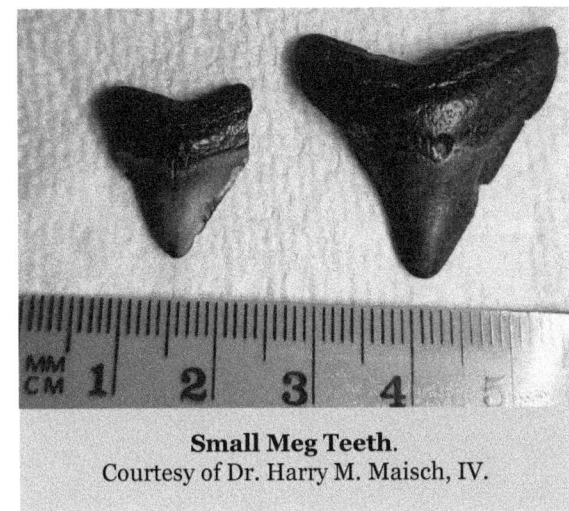

Small Meg Teeth.
Courtesy of Dr. Harry M. Maisch, IV.

Another nursery for copper sharks was discovered in Peru in the late Miocene.[3] Some other locations believed to be Baby Shark daycare centers include Panama, Maryland, Spain, Florida, and South Carolina.[2,4]

The Miocene Gatun Formation in Panama was identified as a nursery for Megalodon, dating back approximately 10 million years, following further analysis of the small-sized teeth found in the area.

Woman Snorkels in Coastal Waters with Young Sharks.
© iStock. Credit: RainervonBrandis.

Young Blue Sharks Along Atlantic Ocean Shore.
© iStock. Credit: Damocean.

Big to Small Megalodon Teeth. Dr. Jay M. Lipoff.

Researchers reported that many belonged to newborn or juvenile Megalodons, measuring anywhere from 6.5 to 35 feet (~2 to 10.5 m) in length.[5] They believe this proves sharks used shallow habitats as a strategy for survival.[7]

Before young Megalodons inhabited the shorelines, *Otodus angustidens* may have had a paleo-nursery during the Oligocene Epoch. Many small neonate, juvenile, and a few larger adult teeth were discovered in the Chandler Bridge Formation, now covering an area once known as the Charleston Embayment near Charleston, South Carolina.[5]

The South Carolina location was previously a shallow, warm, and protected habitat for many prey species and potential predators, which supports the theory that it was an ideal place for a nursery.[4]

Recently, several areas have been reported as nurseries for other living shark species. Hundreds of young great white sharks are cruising up and down the shallow temperate coastline off Long Island, New York.[7] It's time to change vacation plans. However, this is not because there are more humans but more seals.

At Padaro Beach, California, Shark Lab tagged and studied 22 great white sharks between the ages of one and six. They found the sharks hunted in deeper waters at dusk and dawn, stayed near the surface during daylight hours to expedite growth, and were not completely solitary in their younger years.[8,9,10]

Sharks Along Palm Beach, Florida. © iStock. Credit: 6383180.

Interestingly, many coastal regions nearby had similar water temperatures, but that did not draw them in. A constant supply of food or an as-yet-undiscovered resource may play a role in selecting prime nursery areas. A Miocene baleen whale nursery, believed to have existed along the shores of Hiroshima, Japan, could be one such clue.[11]

Within the reefs and shelter of the Galapagos Islands, biologists discovered hammerhead sharks were stopping by to give birth to their newborn and then leaving. Although it sounds like desertion, this South American hideaway is a perfect nursery that offers protection and a bountiful selection of nourishment.[12]

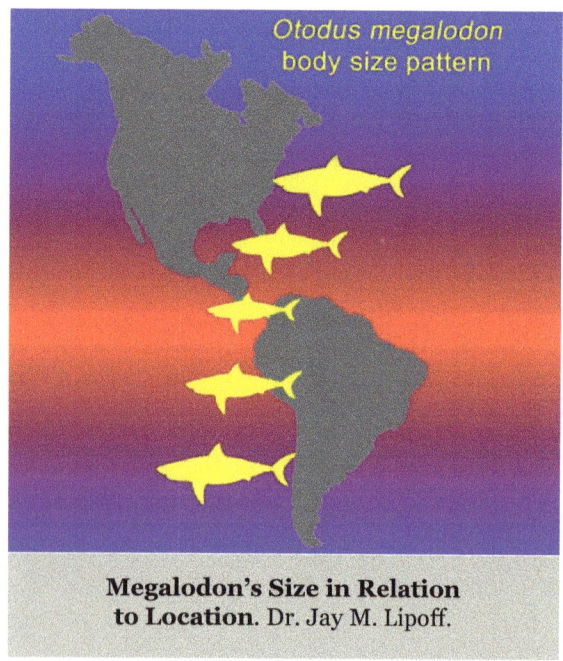

Megalodon's Size in Relation to Location. Dr. Jay M. Lipoff.

In a Natural History Museum article, Dr. Harry M. Maisch IV, from the Department of Marine and Earth Sciences, The Water School, and Florida Gulf Coast University, stated, *"It is still possible that Megalodon could have utilized nursery areas to raise young sharks. But our study shows that fossil localities consisting of smaller Megalodon teeth may instead be a product of individual sharks attaining smaller overall body sizes simply as a result of warmer water."* [13]

In a personal correspondence with him, he emailed, *"The one thing that we all need to keep in mind is that collectors are after the big teeth. Small (juvenile) meg teeth might be overlooked. Additionally, for scientific studies, the specimens that are utilized should be stored in a publicly accessible museum (so others can re-analyze data)."*

"Many museums don't have the smallest, biggest, or rarest specimens—instead, they accumulate 'average' and often fragmentary material that collectors are willing to part with. So, you have preservation bias and collecting bias that can affect these types of studies."

Stack of Teeth.
Courtesy of Dr. Harry M. Maisch, IV.

Megalodon Tooth Compared to Great White.
Courtesy of Japp Roos Art.

CHAPTER FOUR
Megalodon Taxonomy

The first official and documented evidence of Megalodon's existence came from Louis Agassiz,[1,2,3] a Swiss-born American naturalist, teacher, and geologist. In 1835, the species was named *Carcharodon megalodon*.

Scientific research is a collaborative effort, and a continuous evolution of our understanding has challenged the scientific classification of megalodon. In the 1990s, scientists felt the genus *Carcharocles* was more accurate than *Carcharodon and* changed its name to *Carcharocles megalodon*.

Now, you had Megalodon or *Carcharocles megalodon* and a great white shark or *Carcharodon carcharias*. The teeth of Megalodon and great white sharks appeared similar, and many assumptions, such as size and habits, were based on this premise.[4,5,6] Megalodon was considered a prehistoric, larger version of the great white shark.

Carcharocles megalodon is from the Order Lamniformes, which includes mackerel sharks with familiar names like thresher, mako, and great white sharks,[7] as well as other species like megamouth, basking, sand tiger, and goblin sharks.

However, more detailed research comparing great white and Megalodon teeth,[8,9] based on factors including the shark's growth rate and tooth serrations, has revealed a significant shift in our understanding of this majestic hunter's lineage.

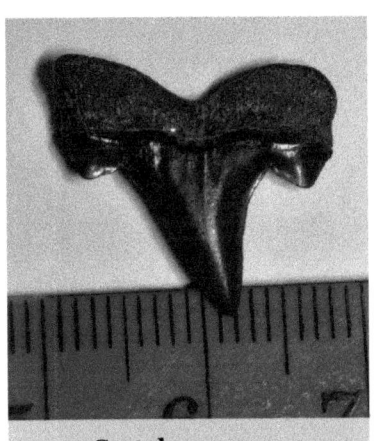

Cretalamna on a cm. ruler. Courtesy of Dr. Harry M. Maisch, IV.

After more research, it is believed that Megalodon's ancestors evolved from sharks of the genus *Cretalamna*[10] at the end of the Cretaceous Period.[11] This placed Megalodon in the family Otodontidae rather than Lamnidae.[12,13,14,15]

Therefore, the presently accepted scientific name for Megalodon is *Otodus megalodon*. As the evolutionary trend of the Megalodon toward gigantism continued, numerous changes occurred in the design of its teeth.[12]

Sixty million years ago, the megatooth shark *Otodus obliquus* had a 3.5-inch (~8.9 cm) tooth with smooth edges and two pointed cusplets on either side of the main tooth crown. It looked like a fork with two small side prongs.

This may have been because it enabled the shark to grasp and tear into faster prey or perhaps to prevent gum disease from occurring in the gaps between the teeth.[16]

During the Eocene, approximately 54 Mya, *Otodus auriculatus* had a slightly larger tooth and smaller cusplets, with both the main crown and cusplets featuring serrated edges.[16]

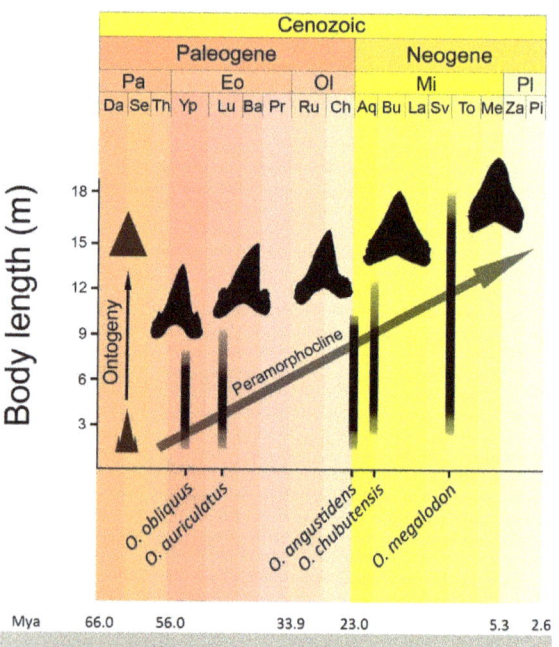

Morphologic Changes in Otodus Teeth Over Time. Ballell, A., & Ferrón, H. G. (2021). Biomechanical insights into the dentition of megatooth sharks (Lamniformes: Otodontidae). *Scientific reports, 11(1)*, 1232. Courtesy of Dr. Humberto G. Ferrón.

Otodus angustidens was an ancestor from the Oligocene, approximately 33 Mya. This species had teeth that grew larger and broader, measuring around 4.5 inches (~11.4 cm) in length, with smaller cusplets and serrated edges. However, the tooth became broader.

It wasn't until 23 Mya in the early Miocene that *Otodus chubutensis* almost completely lost the cusplets, and the tooth slightly increased in size again.[16]

It began to resemble the wide, serrated, thick *Otodus megalodon* tooth from 18 million years ago in the middle Miocene, the one we all dream of finding.

It took roughly 12 million years of evolution to lose the cusplets and finally create the ultimate carnivorous weapon.[16]

At the same time, if you laid these teeth down on their front side and looked up from the root, the *Otodus obliquus* tooth had a large arch.

Through evolution, this arch decreased and became more flattened as it evolved into *Otodus megalodon*.

In a detailed study of the evolution of teeth found in the Calvert Cliffs region of Maryland, Dr. Victor J. Perez noted that, over a 14-million-year time frame, the number of teeth he studied with cusplets consistently decreased.

Between 20 and 17 million years ago, 87% of the teeth recovered had cusplets. This fell to 33% from 16.4 to 14 million years ago; between 10.4 and 7.6 million years ago, they were absent.[17]

When these new visible differences and evolutionary dots were connected, they realized that Megalodon was more closely related to mako sharks than to great white sharks.[14,18]

Great white sharks also have a separate lineage from mako sharks and may have evolved from *Carcharodon hastalis* to *Carcharodon hubbelli* to *Carcharodon carcharias*, the great white.[1]

To summarize the changes in the megatooth shark's teeth over the past 40 million years, the following are the key developments:

1. They lost their lateral cusplets, the smooth edges of the teeth became serrated for cutting, the teeth flattened, and the teeth widened at the crown.[11,15,17]

2. Throughout their evolutionary process, the ability of sharks' teeth to withstand greater bite pressure increased, as seen in the diagram to the right.

   Initially, this dropped during the act of attacking prey by puncture or a vertical bite force,[15] and draw or a horizontal lateral force, as the shark evolved from *O. obliquus* to *O. auriculatus*.[12,15]

   During the next three evolutionary changes from *O. angustidens* to *O. megalodon*, the capability of the front, side, and posterior teeth to handle the greater pressure necessary for hunting rose substantially.[11,12,15] It more than doubled in the front and lateral teeth.

   Essentially, the force was greatest at the tip of the tooth and was then dispersed down to the root equally in teeth shaped like an isosceles triangle.

   However, if the tooth was shaped more like a triangle with a 90-degree straight side, the vertical edge took the brunt of the pressure, as in *O. obliquus, O. auriculatus,* and *O. angustidens.*[12]

3. The Otodus sharks transitioned from grasping faster prey to tearing the flesh off larger prey.

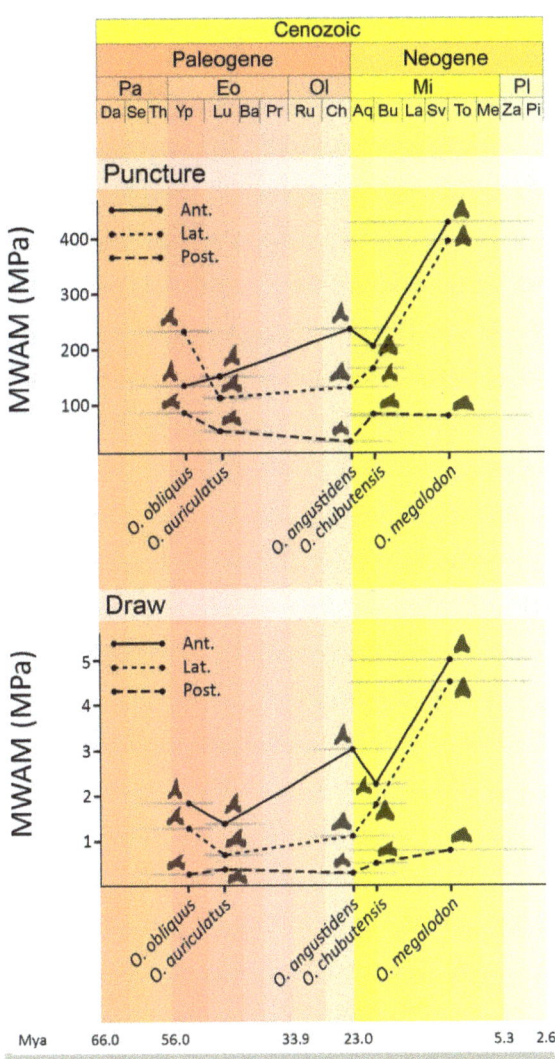

Bite Pressure Changes. Ballell, A., & Ferrón, H. G. (2021). Biomechanical insights into the dentition of megatooth sharks (Lamniformes: Otodontidae). *Scientific reports, 11(1)*, 1232. Courtesy of Dr. Humberto G. Ferrón.

Paleontologist Dr. Victor Perez explained this to me in an email as follows:

"The evolution of the megatooth lineage seems to be closely tied to the evolution of marine mammals. Over a roughly 40-million-year time span, the megatooth lineage became increasingly well-adapted to preying on large marine mammals."

"Over this time, the teeth in this lineage increase in maximum size, acquire an evenly serrated cutting edge, and become wider and flatter proportionally. These features all add up to form a highly efficient, blade-like cutting tool used to dismember large-bodied prey."

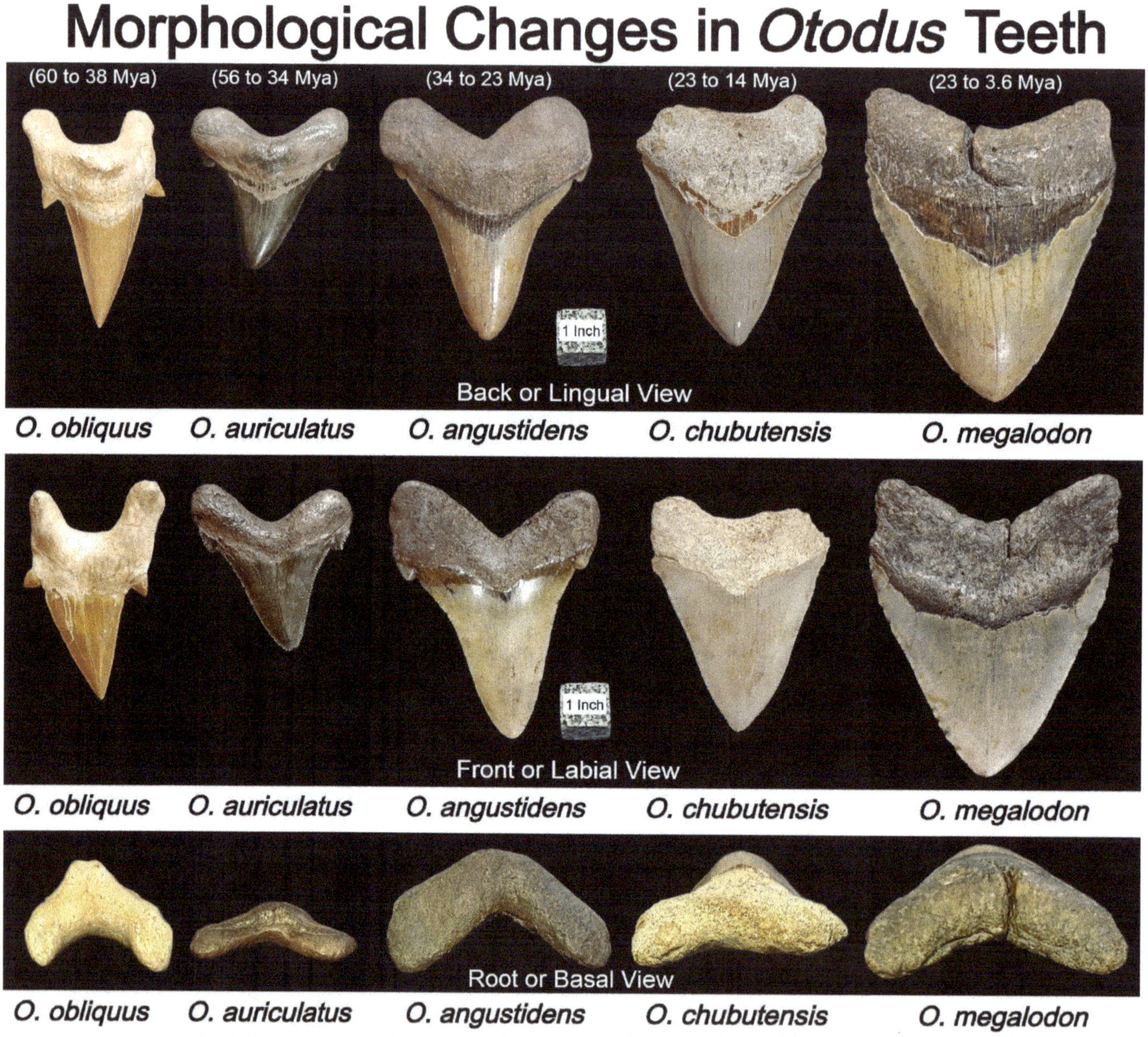

Morphological Changes in *Otodus* Teeth

(60 to 38 Mya)	(56 to 34 Mya)	(34 to 23 Mya)	(23 to 14 Mya)	(23 to 3.6 Mya)

Back or Lingual View

| O. obliquus | O. auriculatus | O. angustidens | O. chubutensis | O. megalodon |

Front or Labial View

| O. obliquus | O. auriculatus | O. angustidens | O. chubutensis | O. megalodon |

Root or Basal View

| O. obliquus | O. auriculatus | O. angustidens | O. chubutensis | O. megalodon |

Photos by Dan Case. Created by Dr. Jay M. Lipoff & Harry M. Maisch, IV, Ph.D.

Morphological Changes in the *Otodus* Lineage

Otodus obliquus

Period – Early Paleocene to Middle Eocene (60 to 38 Mya)

Body Size – Up to 30 feet (~9 m)

Teeth – Could reach 4 inches (~10.2 cm)

Shape – Triangular (isosceles), 3-pronged like a fork, moderately arched root

Cusplets – Yes, large, triangular, smooth edge, separate, pointing outward

Serrations – None

Otodus auriculatus

Period – Early Eocene to Early Oligocene (56 to 34 Mya)

Body Size – Up to 31 feet (~9.5 m)

Teeth – Could reach more than 5 inches (~12.7 cm)

Shape – Triangular (isosceles), similarly shaped

Cusplets – Yes, smaller, rounded tips, irregular serrations, and connected

Serrations – Yes, irregularly spaced on the tooth and cusplets

Otodus angustidens

Period – Oligocene (34 to 23 Mya)

Body Size – Up to 39 feet (~12 m)

Teeth – Up to 4.5 inches (~11.4 cm)

Shape – Triangular (isosceles), serrated edges, mild root arch

Cusplets – Yes, small, rounded with fine serrations, separated from tooth by a visible notch

Serrations – Yes, small, along all edges like a knife

Otodus chubutensis

Period – Late Oligocene to Middle Miocene (23 to 14 Mya)

Body Size – Up to 44 feet (~13.5 m)

Teeth – Up to 5 inches (~12.7 cm)

Shape – Triangular (equilateral), serrated edges, broader tooth base, flatter root

Cusplets – vestigial, triangular, and rounded, with serrations

Serrations – Yes

Otodus megalodon   **Bold** font indicates the latest research

Period – Early Miocene to Early Pliocene (23 to 3.6 Mya)

Body Size – 50 to **80** feet (~15.24 to **24.3** m), 120,000 to **207,000** lbs. (~50,000 to **94,000** kg)

Teeth – Up to over 7 inches (~17.8 cm)

Shape – Triangular (equilateral), broad with a flatter root

Cusplets – Not in adults

Serrations – Yes

©2025 Dr. Jay M. Lipoff and Harry M. Maisch, IV, Ph.D.

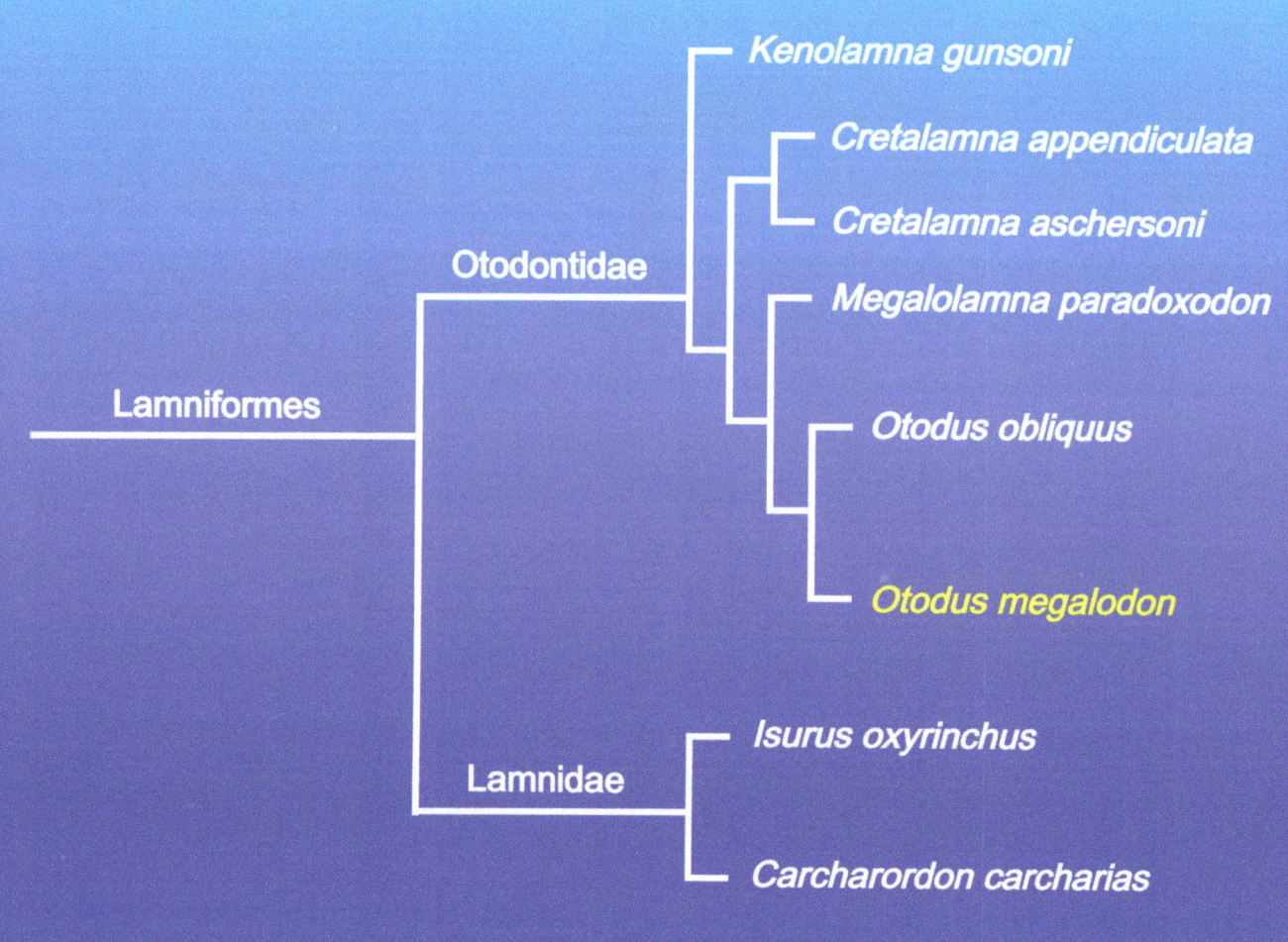

Hypothetical Relationships Between Megalodon and Other Sharks. Dr. Jay M. Lipoff, (2025). Redrawn from information created by Shimada et al. (2016), Ehret et al., (2009), and the findings of Siversson et al. (2015).

This new understanding of Megalodon's evolution opened a world of intriguing possibilities and challenged our previous theories. The key takeaway from this is that Megalodon sharks are not directly related to great white sharks. Therefore, much of what we believe regarding length, girth, diet, and so on may be inaccurate, just like the genealogy of its name. Time will uncover more truths.

The complete classification of the Megalodon is as follows. Its Domain – Eukaryota, Kingdom – Animalia, Phylum – Chordata, Class – Chondrichthyes, Subclass – Elasmobranchii, Subdivision – Selachimorpha, Order – Lamniformes, Family – Otodontidae, Genus – *Otodus*, Species – *Otodus megalodon*.[8]

A new study, recently published in 2025, has taken the fossil world by storm and significantly altered our understanding of Megalodon. Based on the idea that Megalodon's ancestry could be more accurately traced to mako sharks, new recalculations indicate that many studies may have underestimated Megalodon's length.[19,20,21] It may have been longer and leaner than previously thought.[18,22]

In a cross-sectional view, the body would have been more oval than round,[20] further suggesting a slimmed-down and streamlined version of what we used to believe.[22] They may have resembled more of a lemon shark in their appearance.[22] This torpedo-shaped body would enhance its hydrodynamics and efficiency.[22]

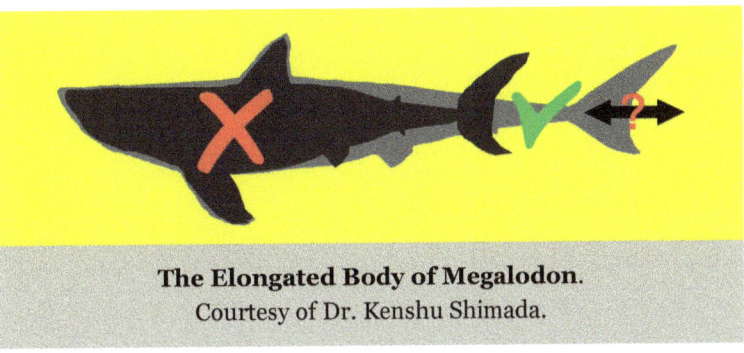

The Elongated Body of Megalodon.
Courtesy of Dr. Kenshu Shimada.

Due to the rarity of complete Megalodon skeletons and the fact that even when one is discovered, it is often incomplete, our understanding of this renowned creature is limited in many aspects.

As stated previously, many theories to date have revolved around extrapolations of tooth sizes as an indicator of length and from an incorrect association with Megalodon's appearance as simply a larger version of the great white shark.

Conventional Shark Design. ©Public Domain. The speculative rendering of this shark was used only to clarify the three sections of the Conventional Shark Design in the study. A 6-foot (~1.83 m) human was added to simulate the size of a shark if it were 80 feet (~24.3 m) long. Created by Dr. Jay M. Lipoff.

To combat this barrier, Dr. Shimada and his team re-examined vertebral fossils from 165 shark species, covering ten different orders.[22] They determined, through a more detailed analysis of the *Otodus* lineage, that the porbeagle shark (*Lamna nasus*) and the lemon shark (*Negaprion brevirostris*) may have had the most similar body part proportions to Megalodon despite having evolved different lifestyles.[22]

They closely studied the dimensions of the three major body sections of sharks:

1. The neurocranium or skull length (NL)
2. Trunk proportion (TP)
3. Trunk-to-caudal or tail-to-fin length (CL).[22]

Together, these three measurements make up the Conventional Shark Design. As evolution experimented with different combinations and created a pathway for maximum survival, these numbers show a wide range of values, reflecting the incredible diversity and morphological changes in Lamniformes.[23,24,25]

Porbeagle Shark (*Lamna nasus*). Courtesy of NOAA.

Livyatan vs Megalodon.
Courtesy of Jaap Roos Art.

CHAPTER FIVE

Megalodon on the Hunt

Apex Predator

There is much mystery surrounding the existence of this apex predator. Sometimes, a single answer can lead to more questions. Imagine the ancient marine ecosystem where Megalodon, the largest macropredatory shark known to have hunted the sea, coexisted with other formidable creatures.

During Megalodon's existence, many species of sperm whales thrived. These toothed whales would have varied in body size. The possibility of these behemoths colliding with one another while focused on their prey is intriguing and captivating.

One such creature was *Livyatan melvillei*, a monstrous sperm whale discovered in Peru[1] and Beaumaris Bay, Australia.[2] Its head was 9.8 feet (~3 m)

Livyatan melvillei **skull**. Photo by Ghedoghedo on Wikimedia Commons. Licensed under CC BY-SA 4.0.

long, and the teeth were approximately 14 inches (~36 cm) long and 4.72 inches (~12 cm) wide.[3,4] Those are some big teeth. The largest of any animal on Earth. Even the *Tyrannosaurus rex*.

The potential for these two behemoths to have battled one another for territory or a meal would have been nothing short of epic[5]—like Godzilla versus King Kong. Fossil evidence of lesions made by a macro-predator on vertebrae and rib fossils suggests that medium to small prey were targeted for active predation.[6,7]

Another formidable opponent may have been squalodon, a 10-foot (~3.3 m) toothed whale that lived roughly between 28 and 14 million years ago in the waters of Europe, Australia, and North America. No fossils have been found in Florida; however, they have been discovered in the Chesapeake Bay region, as well as in North and South Carolina.

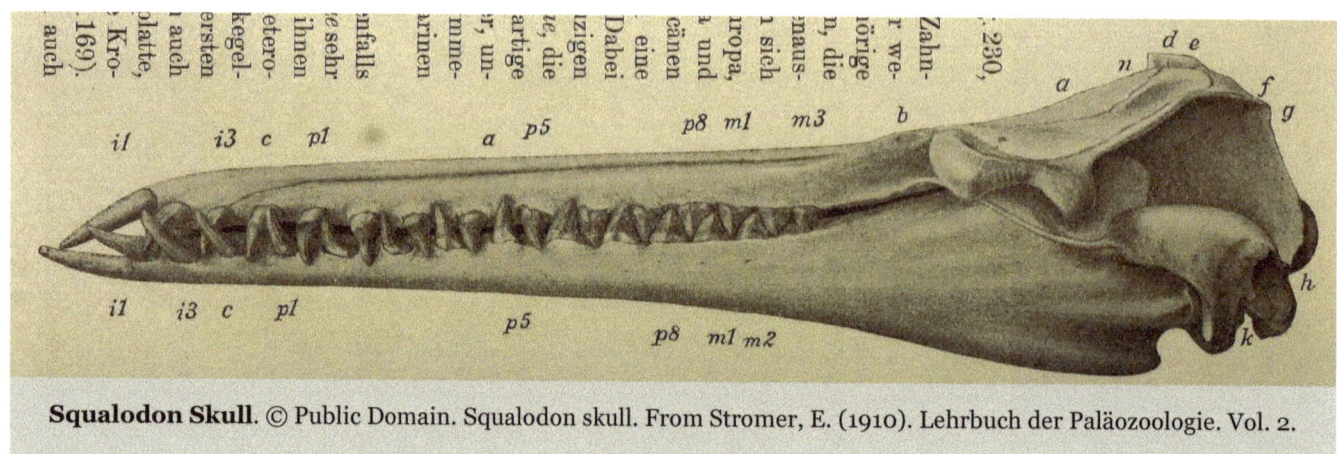

Squalodon Skull. © Public Domain. Squalodon skull. From Stromer, E. (1910). Lehrbuch der Paläozoologie. Vol. 2.

It had tooth features similar to those of a shark and has been referred to as a "shark-toothed dolphin." While only around 10 feet (~3 m) in length, they would be no match for a Megalodon individually. But what if they hunted in packs like orcas?[8] That would be a game-changer. It might have been a predator of juvenile Megalodons and played a role in the balance of nature.

Megalodon likely shared a dietary menu with cetaceans, such as whales and dolphins, as well as pinnipeds, including seals, sea lions, and walruses. There definitely could have been some competition between these apex predators.

Speed

If you have ever caught a fish, you have probably noticed scales along its sides. If you slide your hand along the side of the fish, from head to tail, the scales will lie smoothly and aerodynamically.

The scales become rougher and sharper when you slide your hand up from the tail toward the head. Some scales are completely smooth, while others have tiny ridges or spines running down their surface.

These are placoid scales that cover a fish like a protective layer. They can reduce drag, increase thrust, protect it from bites and abrasions,[1,2,3] and reveal its swimming habits. Due to their different benefits and specialized functionality based on design, sharks strategically have these tiny dermal denticles over their entire bodies for optimal performance and maximum advantage.[3]

A study of the placoid scales that were found near some teeth identified them as belonging to the same Megalodon. However, these clues into the life and times of the Megalodon tell a different story. It wasn't as fast as a white shark or mako.[3,4]

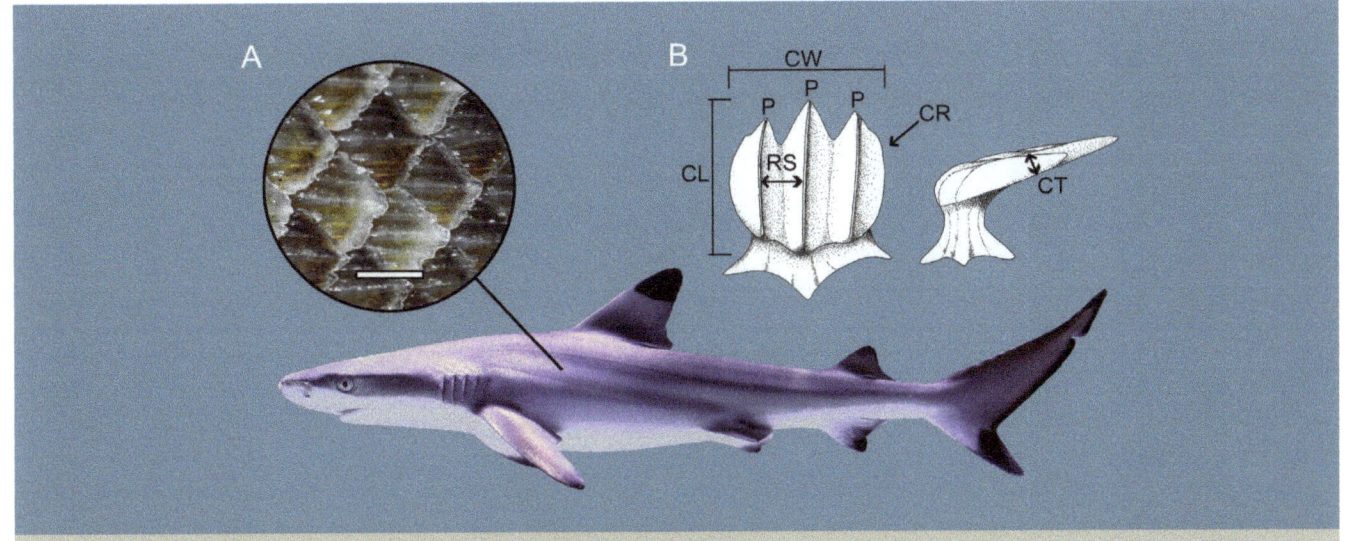

Dermal Denticles of Modern Sharks.
A. Blacktip Reef Shark. **B.** Illustration of dorsal and lateral denticles on a Lemon Shark Body. CR: crown; CL: crown length; CT: crown thickness; CW: crown width; P: peak; RS: ridge spacing. Dillon et al. (2017). Marine Ecology Progress Series. Courtesy of E. M. Dillon.

On Megalodon's placoid scales was a medial spine, also called a keel, similar to the keel on the bottom of a boat. One of its functions is to assist with speed and steering.[5] The keels on the scales of fast-swimming fish are narrowly spaced apart, whereas the keel on a Megalodon scale is widely spaced.[6]

After comparing Megalodon keels and scales against a broad spectrum of fish, researchers determined that Megalodon may not have been a fast swimmer. It may have once cruised along swiftly, and young Megs may have been quick, but after evolving into this immense size, known as gigantism, it shifted into a more leisurely and more efficient swimmer.

Some previous studies suggested that Megalodon cruised at one of the fastest speeds of any oceanic mesothermic animal that ever existed. The belief was that it may have moved along at an average speed of 3.1 mph. (~5 km/h),[7,8,9,10] with an estimated burst speed of 23 mph (~37 km/h).[7,8,9,10]

For comparison, a great white has a short burst of speed, clocking in at close to 35 mph (~56 km/h).

Megalodon Attacking Livyatan. Courtesy of Jaap Roos Art.

***Carcharodon carcharias* or Great White Shark Breaches While Attacking Seal Decoy, South Africa.**
© iStock. Credit: Alessandro De Maddalena.

Utilizing its superpower to regulate the warmth of its locomotive muscles,[8,11] it enhanced its capability to generate the necessary explosiveness to close in, attack, and encounter more prey.[12,13] It also achieved this through improved aerodynamics, resulting in greater efficiency.

Although it cruised at speeds less than extant or current mackerel sharks (lamnids),[1] it potentially was much faster than the whale shark, the largest species in the study.[7]

Attack theories included a burst of speed after their prey, striking when their target was near the surface, possibly breaking ribs, fracturing vertebrae, or biting off a fin to debilitate their prey before the final strike.[14]

It might become scavenger-like and wait for its target to die before attacking, especially with larger targets that could potentially thrash and injure the shark. This is self-preservation at its finest.

On average, 22 teeth on the bottom row pierced the flesh and held it in place. The top of the jaw extended with its 24 serrated steak knife teeth and sliced off severed slabs of meat when the shark shook its head. How much it could eat and how long it could sustain itself on a meal is all speculative.

Based on previous calculations of Megalodon growing up to 50 feet (~15.24 m), it had a stomach as large as 2,600 gallons (~10,000 liters).[7,15,16] Megalodon would tear off and swallow large chunks of flesh, a technique known as lunge-feeding,[17] and then relied on the production of endothermic energy to digest the big meal.[18]

A full-grown Megalodon could eat prey around 9.8 to 19.7 feet (~3 to 6 m) in a few bites.[15] If hungry enough, it might devour an entire orca in five bites,[9,12] roughly 8 to 10,000 pounds (~3,630 to 4,536 kg). That would be about 70% of its estimated capacity,[7] and that was enough. Megalodon could not finish eating prey larger than the modern humpback whale.[7]

Small prey 6.5 to 10 ft. (~2 to 3 m) might have kept it full for 1 to 2 days. A *Metaxytherium*, similar to a 10 ft. (~3 m) manatee, could sustain itself for 2 weeks, and prey over 39 ft. (~12 m) sustained the meg for 2 or more months.[7,9]

This was based on the knowledge that a great white shark, traveling at a slower speed, could explore almost 7,000 miles (~11,265 km) without stopping for food.[19]

However, until we find a live one or unearth a compelling new fossil, **our understanding of current fossil discoveries is based on speculation** and refinement, and theories are always subject to change when we reinvestigate fossil evidence.

For example, we know Megalodon evolved from sharks, but we have no idea what their pectoral, dorsal, or tail fins looked like. Were they big or small, long or short, rounded or pointed, and where exactly were they located? These are all educated guesses. We won't know until we find a more complete fossil or impression.

The latest research, published in Palaeontologia Electronica by Dr. Shimada and 28 colleagues, suggests that if Megalodon had a leaner, more efficiently designed body,[18,20] resembling that of a lemon shark, the amount of food needed and the time required to absorb nutrients would have been dramatically reduced.[18]

At a weight of potentially 200,000 pounds (~94 metric tons), a full-sized 80-foot (~24.3 m) Megalodon may have consumed approximately 3,500 pounds (~1,590 kg) of food daily or every other day.[20] It had an immense appetite and was an apex predator.[20]

This was based on the values of other marine animals and their Fineness Ratio, which is calculated by dividing the creature's length by its profile height.

Fusiform Shape. © Public Domain. NASA.

Marine mammals exhibit a fusiform shape that tapers at both ends, reducing drag. Losing appendages is a further streamlined option, such as swimming forward with your arms out wide versus against your sides, as well as controlling buoyancy, swimming in formation, and riding waves.[21]

The ideal Fineness Ratio for minimal drag is between 3.0 and 7.0, with an optimal value of 4.5, which is where dolphins rank. A great white shark is estimated to be between 4.5 and 5.0 meters in length. These values increase slowly as a species' mass and size increase.

This is why we have only discovered large, great white sharks around 20 feet (~6.1 m) long. It maximizes their success, whereas becoming too large would create hydrodynamic disadvantages, in theory, and make them less successful.[20]

Most large migratory whales scored values ranging from 6.1 to as high as 8.0.[22] Megalodon's Fineness Ratio was approximately 6.08, meaning it was not as chunky as the great white but had a slender body, which allowed it to be more efficient through the water by blending speed with energy conservation.

The study's published cruising speed for a Megalodon is estimated to be between 1.3 and 2.2 mph (~2.1 and 3.5 km/h), with a maximal speed of around 11 and 19 mph (~18 and 30 km/h).[20] This is a slight variation from previous estimates. Refer to the chart below for a side-by-side comparison of the most recent study findings on Megalodon.

Comparison of Previous and Current Megalodon Traits © 2025. Dr. Jay M. Lipoff.		
Traits	**Previous Megalodon Findings**	**Newest Megalodon Findings**
Visual Comparison	Similar to Great White Shark	Similar to Lemon Shark
Size	50 to 60 feet (~15.24 to 18.3 m)	Up to 80 feet (~24.3 m)
Weight	Around 132,000 to (~60 metric tons)	Around 207,000 pounds (~94 metric tons)
Vertebrae	Up to 6 inches (~15.5 cm) Diameter	Up to 9 inches (~23 cm) Diameter
Cruising Speed	1.3 to 2.2 mph (2.1 to 3.5 kmh)	Up to 3.1 mph (~5.0 kmh)
Maximum Speed	Around 23 mph (~37 km/h)	Between 11 to 19 mph (~18 to 30 kmh)
Attack	Ambush With Short Bursts of Speed	Ambush With Short Bursts of Speed
Body Shape	Round	Oval
Appearance	Bulky	Lean, Stream-lined, Elongated
Food	2,500 pounds (~1,134 kg) per day	3,500 pounds (~1,588 kg) per day
Frequency of Meals	Every Few Days	Daily or Every Other Day
Thermophysiology	Endothermic	Endothermic
Range and Distribution	Tropical to Temperate Waters	Tropical to Temperate Waters
Size at Birth	6.5 to 10 feet (~2 to 3 m)	12 to 13 feet (~3.6 to 3.9 m)
Habitat	Coastal Habitat or maybe Open Waters	Open Waters or maybe Coastal Habitat

Global Traveler

Scientists now believe that Megalodon was a warm-blooded shark,[1] which would help explain its size, territory, and hunting grounds. If an animal is endothermic, it absorbs energy from its surroundings in the form of heat and can elevate its body temperature[2] as needed.

Endothermy can also be affected by nutrient absorption and food processing. Sharks have a spiral-shaped intestine,[3] which creates a larger surface area,[4,5] allowing more nutrient absorption from ingested meals over a longer period. The intestinal system and the large, lipid-rich liver are two of the warmest organs in the shark.[6]

An elongated body cavity may also have contributed to the efficiency of heat drawn from its food. This has been the case in other species, and these internal furnaces may have enabled Megalodons to forage for food worldwide, as evidenced by global fossil findings.[7]

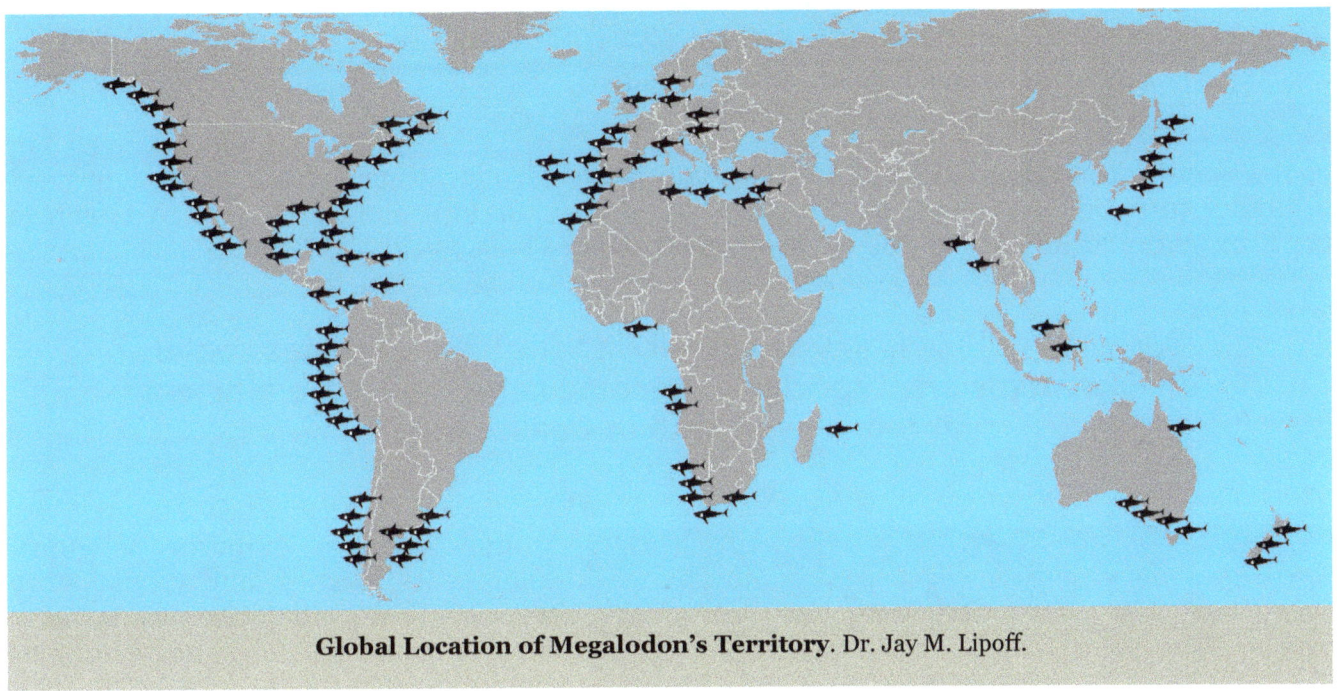

Global Location of Megalodon's Territory. Dr. Jay M. Lipoff.

More research was done on Megalodon teeth and revealed that the shark may have been around 12.5 degrees Fahrenheit (~7 degrees Celsius) warmer[8] than many of the oceans they swam in and consistently maintained a core temperature of 80 °F (~27 °C).[9,10]

The shortfin mako, great white shark, and salmon shark share this ability, but can stay even hotter.[11] However, the caloric demands to sustain a high metabolic rate and high body temperature in a big body made Megalodon potentially more vulnerable to the threat of extinction than other sharks.[9]

The only locations where they haven't found evidence of Megalodons are the super-cold regions, such as the polar caps of Antarctica. Can you blame them? It's cold.

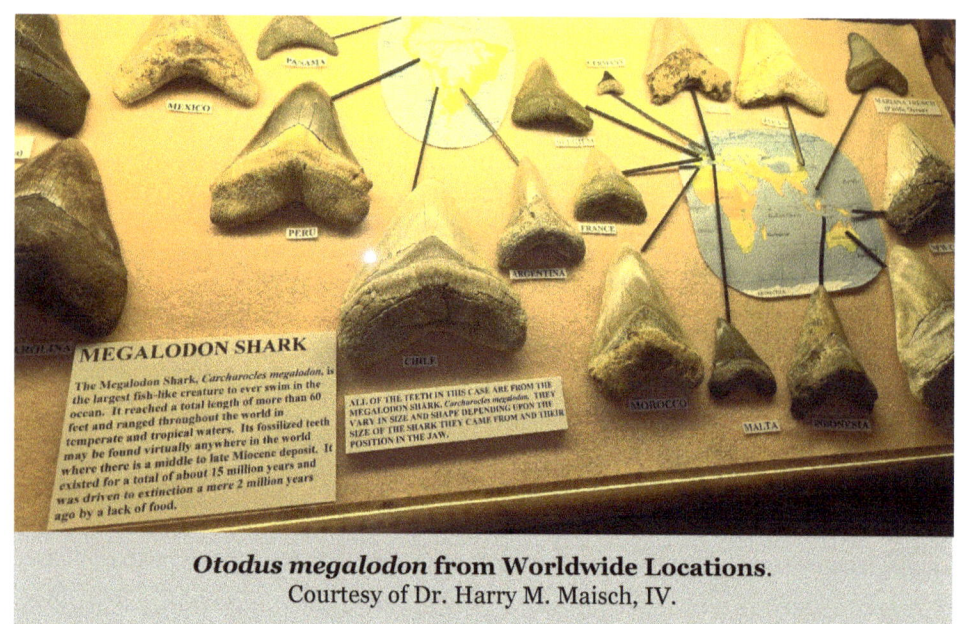

Otodus megalodon from Worldwide Locations.
Courtesy of Dr. Harry M. Maisch, IV.

However, Megalodons did cover much of the globe[2,12] in their search for food,[13] like great whites, and the colder waters did not stop them. They found a tooth in Denmark.[14]

This would have meant that Megalodons were definitely out and about. They went where the food was, which is why Megalodon earned its frequent swimming miles as a transoceanic first-class super traveler.

Megalodons thrived in cooler waters. You become less picky when your stomach's growling because you might need around 2,500 pounds (~1,134 kg)[14] to 3,500 pounds (~1,588 kg)[15] of food, that's 2 to 3 average cows, or 98,000 calories (~12,700 g)[16] to 137,000 calories (~17,755 g).[15]

Bergmann's Rule is a broad generalization[17,18] that states **cooler environments drive species to become larger** and more efficient at retaining heat than smaller ones.

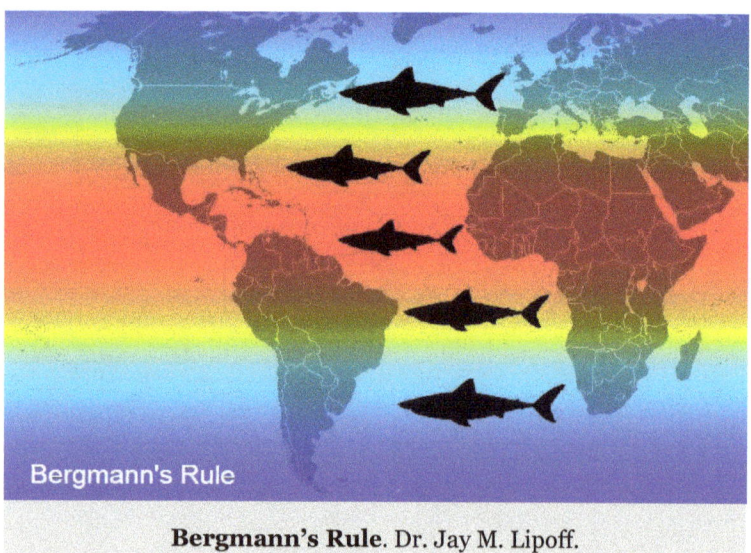

Bergmann's Rule. Dr. Jay M. Lipoff.

Animals such as emperor penguins, polar bears, whales, and moose store energy and accumulate body fat as a form of insulation. They are examples that follow this rule and are larger than their southern counterparts.

Researchers have found that many of the largest Megalodon teeth have been discovered in cooler waters of the higher latitudes, supporting this rule.[18]

Their geographical expansion increased during the Miocene, and they reached their maximum distribution in the late Miocene.

Their sweet spot was between 55.28° N and 43.99° S latitude,[19] roughly from Denmark to New Zealand. Tracking current data, this is where prey, like today's dolphins, whales, and turtles, survive.

Water temperatures in their territory had an annual range of 53.6 °F to 80.6 °F (~12 °C to 27 °C),[19] with a minimum of 33.8 °F (~1 °C) and a maximum of 91.4 °F (~33 °C).[19] However, this was also the beginning of their global population decline.[19]

Like current mako and great white sharks, Megalodon could elevate the temperature of various body parts as needed.[9,10,20] Possessing the ability to regulate body temperature regionally may have been the spark that led to these sharks' gigantism.[20,21]

Essentially, this was the power to become big and travel wherever they wanted, so they took their giant-sized bodies worldwide.[10,22]

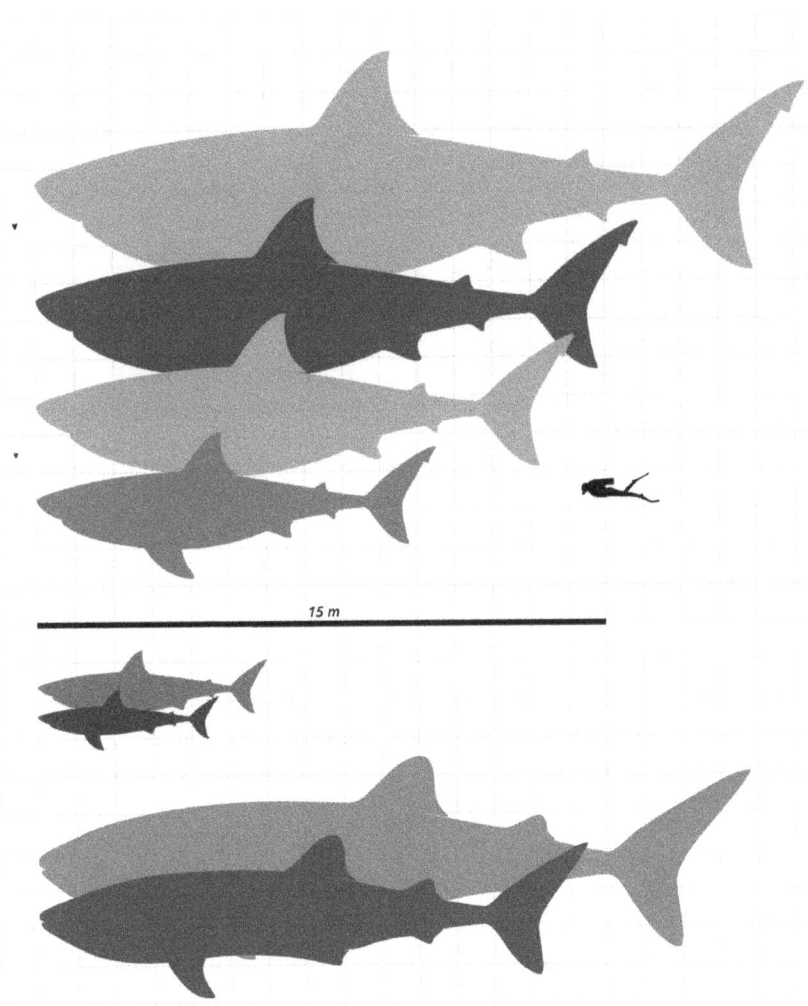

Otodus megalodon

Estimate from tooth GHC 6 using the summed crown width method (SCW) - Perez et al. (2021)	Mean ~20.3 m (~17.4 - 24.2 m) Mean ~ 66.6 ft (~57.1 - 79.4 ft)
Reconstruction of vertebral column IRSNB P 9893 - Cooper et al. (2022)	~15.9 m ~52.2 ft
Estimate from tooth NSM PV-19896 using upper anterior tooth crown height equations - Shimada (2019)	~14.2 m ~46.6 ft
Average estimate of 544 teeth using tooth crown height equations. - Pimienti & Balk (2015)	~10.5 m ~34.5 ft

Great White Shark
(Carcharodon carcharias)

Large Female	~19.7 ft or ~6.0 m
Small Mature Female - Gottfried et al. (1996)	~15.4 ft or ~4.7 m

Whale Shark
(Rhincodon typus)

Large Female - Borrell et al. (2011)	~49.2 ft SL(~61.7 ft est. TL) or 15 m SL (~18.8 m est. TL)
Average Fully Grown Female ? - Meekan et al. (2020)	~14.5 m TL ~47.6 ft TL

15 m

Previous Size Comparison of Megalodon, Great White, and Whale Sharks.
Steveoc 86 (bit.ly/MegalodonCarcharodonScaleChart), Removed *Carcharocles megalodon* and replaced it with *Otodus megalodon* name, added standard American Imperial measurements, by Dr. Jay M. Lipoff, creativecommons.org/licenses/by-sa/4.0/legalcode.

Megalodon Hunting. Cooper, J. A. et al. (2022). The extinct shark *Otodus megalodon* was a transoceanic superpredator: Inferences from 3D modeling. *Sci. Adv. 8, eabm9424*. Courtesy of Dr. Cataline Pimiento and paleo artist Juan Giraldo.

CHAPTER SIX

Fossil Evidence of Attacks

Did Megalodon hunt alone or together? Studies have shown that some great whites will team up to form non-random social connections[1,2] for patrolling or hunting,[3,4] exchange kill or prey location,[4,5] have different strategies for deep water,[6,7] surface and ambush attacks,[8,9] and have a hierarchy in their social network position.[5,10] Whether Megalodon utilized this strategy is unknown at this time.

The Calvert Marine Museum on Solomons Island, Maryland, has a marvelous find of two vertebrae with compression fractures. These fractures suggest that they had been exposed to a tremendous impact or force,[11] squeezing and compressing them with unimaginable pressure.

The famed Calvert Cliffs were the location of the find, and Dr. Stephen Godfrey, the curator of paleontology at the museum, was ecstatic. The vertebrae had traces of bone growth, so it could have been the prey's lucky day. It must have survived long enough for the body to try to fix the damage.[12]

After analyzing the possibility of infection or spasms being the source of the fracture, it seemed unlikely. Instead, perhaps it was like a scene from "The Meg" movies.

He envisioned a suspenseful explanation reminiscent of an ambush strike emerging from the depths of darkness below—the piercing of flesh and crunching of bones, the thrashing of two behemoths in a struggle for survival, a moment of hopelessness and acceptance of defeat, and then one final desperate attempt against impossible odds by a determined baleen whale for an impossible escape from the clutches of death.

Something less dramatic could have happened, but let's stick with the incredibly plausible. The bottom line is it lived to swim another day. For how long? We may never know, but how cool is it that a large *Otodus megalodon* tooth was discovered alongside the fossils? Guilty as charged or circumstantial evidence?

Deep Groves in Vertebra from a Megalodon Attack.
Courtesy of Dr. Stephen J. Godfrey.

Prehistoric Dolphins Attacked by a Massive Megalodon.
Artist: Tim Scheirer. Courtesy of Dr. Stephen J. Godfrey.

Three other caudal or tail vertebrae of a whale or dolphin, two collected along Bayfront Park (Brownie's Beach) and Willows Beach near Calvert Cliffs, Maryland, and another near the phosphate mine in Aurora, North Carolina, also displayed amazing predation marks. Still, there were no signs of healing this time.[13]

The vertebra shown above is one example from this study. This is a *"lateral view of one of the meg-bitten Miocene fossil dolphin caudal vertebrae. All diagonal gouges were made by one Megalodon tooth,"* according to co-author Dr. Godfrey.

The longest gouge is 1.46 inches (~37 mm), and the animal, possibly a Xiphiacetus, was approximately 11.5 feet (~3.5 m) in length. There is prior evidence that extant or currently living sharks typically attack smaller prey and scavenge larger ones. Megalodon may have focused on pinnipeds, such as walruses, sea lions, and seals.[14,15]

Great whites have been known to attack a dolphin from behind to prevent their approach from being detected,[13,16] with the strategy of biting their tail to disable it.[10] This fossil evidence is similar, although examples of Megalodon's feeding habits are rare.[11,13,14,16,17]

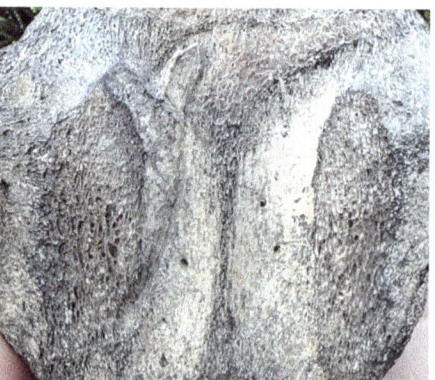

Megalodon Tooth Embedded in Spinal Canal of Whale Vertebra with Healing of the Bone.
Courtesy of Zachary L Coverstone.

While we still don't know how the great Megalodon lived or if these marks were from an attack or scavenger moment, one vertebra confirmed the markings and lasting proof it was wedged between two large adjacent teeth of an *Otodus megalodon* or *Otodus chubutensis*.[11]

There were repetitive bite marks, suggesting that attempts to break free during this struggle for survival were futile. This lends support to the theory that Megalodon was an active predator.[11]

Coincidentally, lightning struck two more times. Dr. Godfrey has also written about a whale rib found in South Carolina that became lodged between two teeth, similar to the vertebra on the previous page.

Dugong Rib with Megalodon Bite Marks.
Courtesy of Ryan Meyer.

The grooves, created by the serrations along the edge of a Megalodon tooth, are a distinctive pattern of Megalodon predation and are forever captured in the fossil.[18]

Another rib fossil from North Carolina was bitten but showed signs of healing, indicating that the whale had escaped death, albeit temporarily. Infection and bone repair stopped on the rib, indicating that the whale didn't survive.[16] However, it does support the theory that the whale was being actively hunted, rather than scavenged.[17]

Fossil finds from the Miocene deposits in Southern Peru include the skull bones of a baleen whale, *Piscobalaena nana*, and pinnipeds. Once again, these bite marks are credited to the huge teeth of an apex predator and megatooth shark, *Otodus megalodon*.[14]

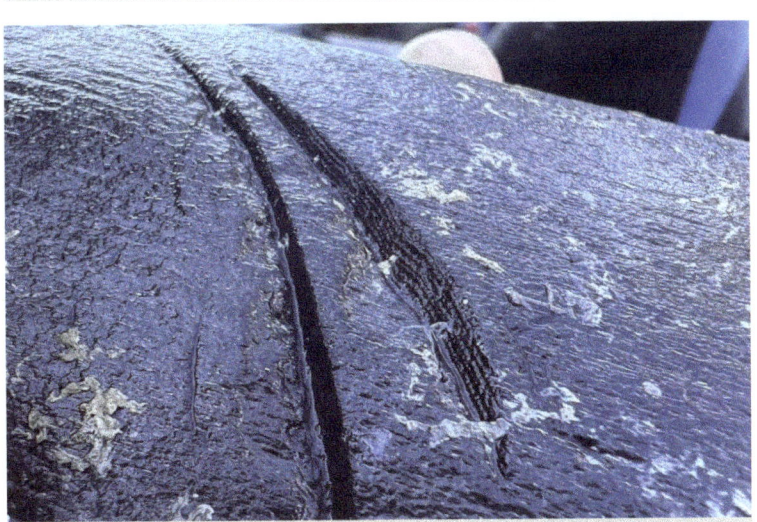

**Serration Marks and Deep Grooves
from a Megalodon Bite in a Whale Rib.**
Courtesy of Frank Morrow. Picture by Blair Morrow.

In the golden days, fossil hunters could explore the phosphate mines in Aurora, North Carolina, and find many large Megalodon teeth and other treasures. Some were turned into museums, like a sperm whale tooth with three gouges, which, from the size, led to the conclusion they belonged to the extinct *Otodus chubutensis* or *Otodus megalodon*.[19]

Still, with serrations from bite marks embedded in the tooth's root, researchers believe this was an attack on the whale rather than a scavenger encounter, as there is not much meat on a skull.[19]

Lamniform sharks, such as Megalodon, have a protruding upper jaw, so a typical bite would not damage another tooth. However, a tooth might dislodge during an attack and sustain a bite mark in the commotion.[20] You will not often find one of these, but they exist.

Sharks do attack each other, too, and for various reasons. Many sharks can be seen with scars from basic survival, mating challenges, and rituals, as well as defending their territory, demonstrating dominance, and more.

It is safe to assume that during Megalodon's existence, it faced many of the same difficulties that today's sharks do, except for pollution and hunting by humans.

Like Marvel's superhero, Wolverine, **sharks have an amazing ability to heal themselves**. Whale, mako, and great white sharks are just a few examples of sharks with incredible genes that enable them to exhibit a high capacity for muscle regeneration,[21] wound healing and recovery,[22,23] skin repair,[21] and the regrowth of damaged fins,[24] particularly those that have not been completely removed.

However, they can get cancer[25] and other diseases,[26] so humans should stop killing or maiming them with high hopes of cancer prevention or virility. There isn't enough evidence to suggest that sharks don't get cancer, but a study indicates that a gene in mako sharks may suppress human tumors.[22,27,28] This implies that they are more useful alive, offering a health benefit to humans, potentially, but also to rule the oceans, as they have for hundreds of millions of years.

The shark species needs to be protected and not driven to extinction. Let's have real scientists research them properly, thoroughly, and slowly to determine if they hold a clue to help humans fight off certain ailments. However, there's no reason you can't continue collecting those monster teeth and other fossils alongside them.

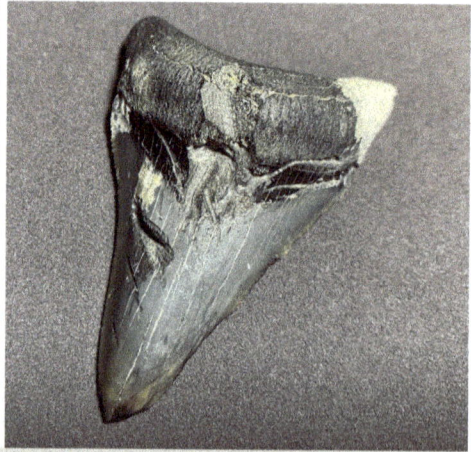

Bite Marks on the Meg Tooth. Courtesy of Blair Morrow.
(Meg Goddess Designs, Aquanutz Scuba Diving Charters).

You'd Better Get Started.
A Friend's Cabinet. (One of Six.) Dr. Jay M. Lipoff.

Heartbreakers.
Courtesy of Blair Morrow and Michael Konecnik
(Meg Goddess Designs and Aquanutz Scuba Diving Charters).

CHAPTER SEVEN

Megalodon Teeth

Otodus megalodon means ear-shaped, large, or giant tooth in Ancient Greek. A Megalodon tooth can be three times larger than a great white tooth. The broad upper teeth, lined with serrations, were to slice, while the lower teeth were narrow, serrated, and designed for piercing prey. Together, they were nature's perfect prehistoric cutlery set for feasting.

It is quite a thrill when you stroll up to a large fossil tooth on an eastern shore or venture off for a dive in the Gulf Coast waters off Venice, Florida, a city often referred to as "The Shark Tooth Capital of the World."

This is because numerous Megalodon teeth and other remarkable fossils have been discovered along the shallow coastline. Florida is rich in fossils from the Miocene and Pliocene epochs.

Upper Tooth of Great White & Megalodon (left);
Lower Tooth of Great White & Megalodon (right).
Courtesy of Dr. Harry M. Maisch, IV.

With all this hype, you might think they would have declared Megalodon teeth as Florida's state fossil. Sorry. According to the North Carolina General Assembly Session Law 2013-189, HB-830, that honor belongs to North Carolina.[1] Also, a worthy candidate for the title, nonetheless.

A good day is finding a shark's tooth of any size while secretly wishing for something to fit into your palm, like a tooth 4 inches from tip to corner. On rare occasions, scientists and hobbyists have discovered a 6-inch tooth, something to celebrate like a hole-in-one golf shot.

Articulated Megalodon Teeth. Collected by the Calvert Marine Museum staff over 6 years. Dr. Jay M. Lipoff.

Regardless of size, they are wonderful treasures. Some may exhibit damage from hunting or have been repurposed by Paleo-Indians, shaped to function as lance points, cutting edges, or pendants.

This can happen at any geologically correct beach,[2] the Calvert Cliffs in Maryland, the Low Country region, and the Cooper River of South Carolina, the submerged continental shelf of Onslow Bay, North Carolina,[3] Peru, and many other locations.

Another Florida hotspot for large teeth and various fossils is Bone Valley in the phosphate mining region of central Florida. In fact, 95 associated teeth from one Megalodon were found in a phosphate mine in Polk County. Countless other isolated teeth have been collected from the Peace River in Florida, which bisects the same region.

In October 2024, the Calvert Marine Museum (CMM) announced it had struck gold, too. Paleontology Collections Manager John Nance was drawn to the sound of a baby deer crying from the base of the cliff while its mother stood helplessly above. When he approached the deer, he noticed a Megalodon tooth. In a moment, another one was visible, and then another. I would guess his heart started pounding in pure excitement.

He contacted Dr. Stephen Godfrey, the curator of paleontology, and Dr. Victor Perez, who was then an assistant at CMM. Over the course of six years, their team carefully recovered 53 teeth, ranging in size from 0.5 to 5.5 inches (~1.3 to 14 cm). Equally important, the Maryland Department of Natural Resources (DNR) reunited the fawn with its mother.

However, it is rumored that another enthusiast found 27 from this same location in 2015-2016. They are attempting to locate them due to their scientific significance and rarity. This is the unfortunate side of fossil collecting.

What would you do if you found 20 Megalodon teeth? Celebrate and share on social media, add them to your collection, sell them, or consider donating them to a museum. In this case, the Calvert Marine Museum would at least like to cast them to make their display more complete.

Variations in Shark Teeth Designs. Teeth from various species and from several locations. Dr. Jay M. Lipoff.

With advances in 3D scanning and printing technology, many museums can now incorporate casts of rare specimens for display. If collectors are willing to donate a rare find, they should be able to receive a scaled replica in return.

Shark teeth exhibit a range of shapes, angles, thicknesses, curves, sizes, and original colors; however, why do their colors vary when fossilized?

Like a human tooth, the enamel of a Megalodon tooth is dense, polished, and smooth, with a hardness of around 5 on the Mohs hardness scale. In contrast, the root and majority of the tooth are composed of rough and porous dentine, which is loosely held to the jaw tissue.

Enamel covers the tooth and is primarily made of hydroxyapatite. It provides increased structural strength, making the teeth stronger than bone. It takes approximately 10,000 years for the calcium phosphate in teeth to be replaced by the surrounding sediments.

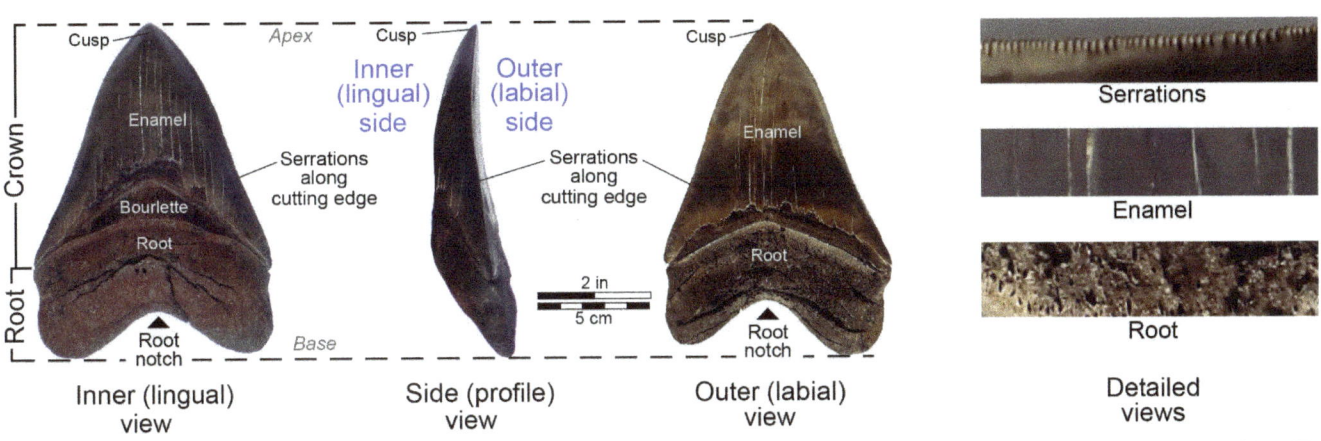

Anatomy of a Megalodon Tooth. Greb, S. *"Fossil of the month: Megalodon teeth."* Kentucky Geological Survey, Earth Resources-Our Common Wealth. Fossil of the Month: Megalodon teeth (uky.edu). Courtesy of Mary Ann Russell, Kentucky Geological Survey Collection.

The cavities and original bone material in the bone are filled with minerals transported by water into these pores as part of the fossilization process, which increases the fossil's weight and can change the color of the tooth.

So far, the largest tooth ever recovered is 7.48 inches (~18.99 cm) and was found in Peru.[4,5] That could indicate that Megalodon and Livyatan swam into one another occasionally. Scientists extrapolate calculations from these giant teeth to determine the shark's size,[6,7] typically by comparing the tooth's crown height to its body length.[8]

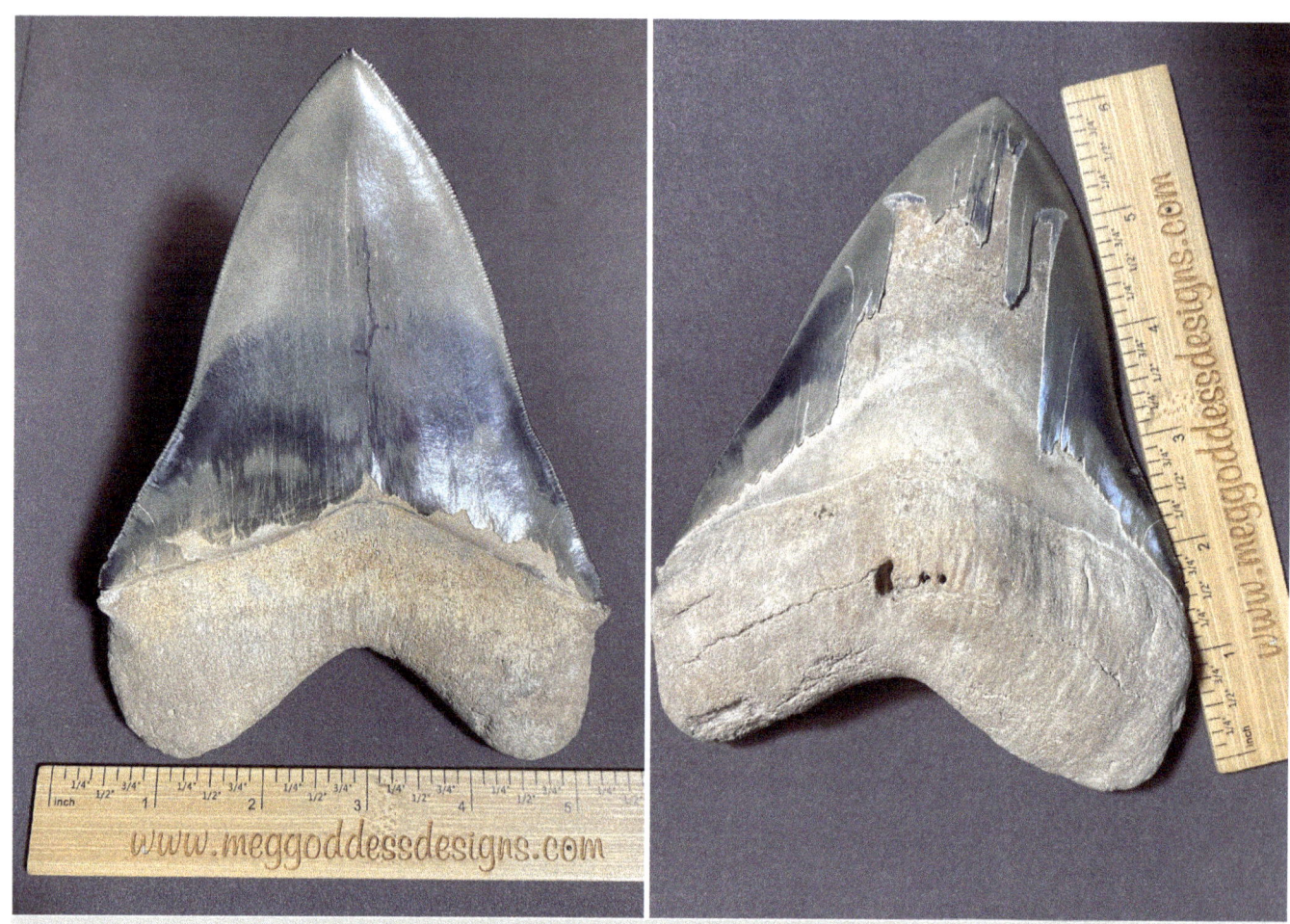

Megalodon Tooth from Florida. Enamel peeling on the back of an over 6" tooth.
Courtesy of Blair Morrow (Meg Goddess Designs, Aquanutz Scuba Diving Charters).

A 7-inch (~17.78 cm) tooth could belong to a shark of 60 feet (~18.3 m) or more.[8] Interestingly, they also needed to consider the tooth's location in the mouth because it is shaped differently and could alter the overall body length estimates.[9,10]

Teeth are measured from the edge of the root to the tip of the crown. For instance, as you move backward, the teeth are curved in that direction to hold their catch, while in the front, they are straight, so they measure longer.

The old theory was that every inch of a tooth equals 10 feet (~3 meters) in length, but this is akin to explaining how, in a dog's life, 1 year equals 7 years of human life. A 2025 study by Shimada et al suggests that this method of estimating Megalodon's length could be drastically inaccurate.[11]

An obvious reason is that if the same shark has big teeth in the front and smaller ones in the back with a slight curve, two teeth from the same shark could yield two different overall lengths.

There seemed to be a never-ending supply of teeth being found because Megalodon had five or more rows of them, and they could continually be produced and replaced. This masterful tooth replacement system evolved to perfection over 400 million years.[12,13]

Their teeth were not in sockets like ours. Specialized cartilage in their jaws allowed these amazing teeth to form continuously. The developing teeth adhered to the gum's connective tissue until needed. Almost 300 teeth might have existed in the rows and rows of teeth that filled Megalodon's jaws at any given time,[14,15] as they awaited their call to action.

Some shark species can replace their teeth every two weeks.[16] As one Megalodon tooth fell out, a new one

Lower Jaw of Fossilized Pliocene Great White Shark Ancestor, *C. hastalis*. Calvert Marine Museum. The teeth roll out of a triangular-shaped cavity. Dr. Jay M. Lipoff.

slid forward on this conveyor belt of carnage, like removing a soda can from the shelf of a gas station vending machine. The gum tissue that once held the tooth was then reabsorbed or disintegrated.

Normal replacement occurred as new teeth from below matured. The old ones were pushed out, and then these took their place. If a tooth were old or broken during feeding, it would eventually be discarded, enabling the shark to swim on with new teeth.

You will notice something interesting in the articulated teeth from the Megalodon exhibit at the Calvert Marine Museum in Maryland.

File Teeth. These teeth are from the articulated Megalodon at the Calvert Marine Museum Dr. Jay M. Lipoff.

The teeth at the bottom of each vertical row, known as file teeth, have a flat root that slowly forms into an arch as they move forward. File teeth are the rows of teeth waiting to rotate to the top.

One shark might use and lose up to 40,000 teeth during its lifetime.[16,17] That means tens of thousands of teeth are left along the ocean floor by a sea once filled with Megalodons.[18]

They are everywhere. What are you waiting for? The ocean is not running out of them.

Like Forrest Gump said, *"You never know what you're gonna get."* Your dive might be filled with heartbreak, like the teeth at the beginning of this chapter, the beauty could be hidden, things could line up nicely, or you might be misled.

David Ousley, who found a 6" plus tooth. He invented underwater bowling, and he's undefeated.
Courtesy of Blair Morrow and Michael Konecnik (Meg Goddess Designs and Aquanutz Scuba Diving Charters).

A July 4th Picture of Some Patriotic Red, White, and Blue Megalodon Teeth. Courtesy of Sven Reiter.

As if the Shark wasn't Scary Enough, Even the Teeth Have a Scary Face. Courtesy of Sven Reiter.

CHAPTER EIGHT

Variations in Megalodon Teeth

Only some of the teeth you find will be suitable for a museum or worth thousands of dollars. Some teeth have traveled great distances or have been exposed on the seafloor, where they were tumbled with rocks, shells, and sand over millions of years.

As a result, they might appear dull, have smooth edges, have freckles or colored spots, have enamel peeling or vertical cracks, have barnacles growing on them, or, better yet, have chips from predation damage from when they ate.

Other unique features include holes created by endolithic bivalves, also known as rock-boring clams. These holes are oval, circular, club-shaped, or dumbbell-shaped, varying in width and depth, and are classified as trace fossils identified as *Gastrochaenolites*.[1]

Clionaid sponges, algae, bacteria, fungi, and serpulid worms can also form borings into the teeth of Megalodon.

These types of bioerosion are frequently associated with lag deposits, which are fossil concentrations that develop in response to fluctuating sea levels and movements of the ocean bottom due to storm currents and erosion.[1]

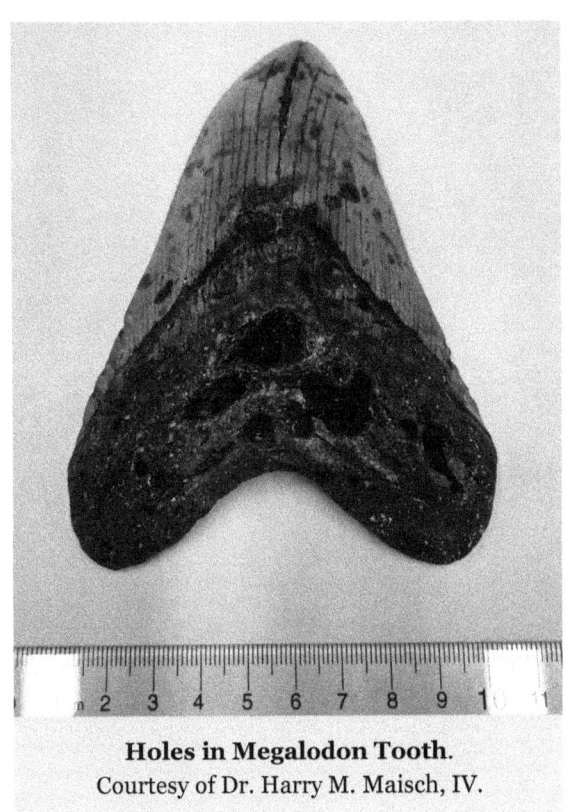

Holes in Megalodon Tooth.
Courtesy of Dr. Harry M. Maisch, IV.

During the mineralization process, a fossil is exposed to various factors that can influence its color. These include the time it takes to become fossilized, heat, the density of the surrounding material, pH, pressure, and chemical interactions. **Because different minerals produce different effects, a fossil could retain many fantastic colors** if exposed to various sediments multiple times or under the right circumstances.

Phosphate mines in Aurora, N.C., or within the shallows of Florida's coastlines and mainland produce jet-black teeth due to dark minerals like manganese. However, the teeth can range from tan to gray or black. The different sediment and rock types will provide varying degrees of permeability and percolation, allowing certain minerals to dissolve and precipitate out.

In South Carolina and Peru, the presence of more iron causes the teeth to take on an orange or reddish-brown color. Tannin from decayed vegetation can cause the teeth to be a varying shade of red or brown, like the color of a stream. Black river coloring happens because soil enters the water as the river absorbs organic acids from decaying plants, like a moving stream of hot tea.

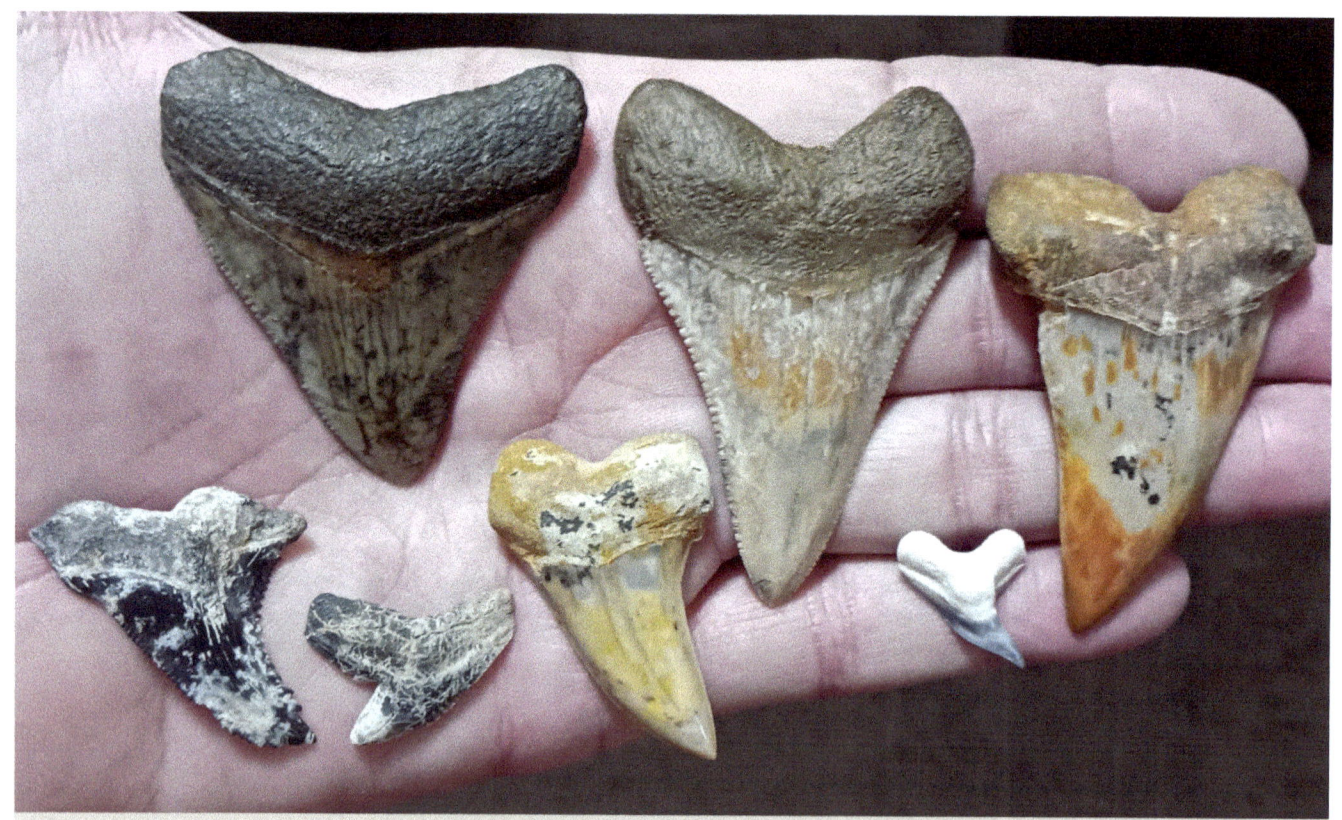

Example of Colored Teeth from Various Species.
Top left to right: *Otodus megalodon, Carcharodon carcharias, Carcharodon hastalis*. Bottom: *Hemipristis serra, Galeocerdo cuvier, Carcharodon planus, Physogaleus contortus*. Courtesy of Dan Case.

Lastly, there is the white tooth. It's not a fresh tooth from a live Megalodon—we can dream—but it occurs when groundwater pulls minerals out of the tooth, leaving it pale or almost white in appearance. Nature's own whitening system.

When a tooth is exposed to the sun or groundwater for an extended period, it can lose its vibrancy and become dull. Environmental factors can add their own artistic touch, such as dark or yellowish speckles, or even prettier yet.

If a tooth has a mottled appearance with red, orange, and black spots, it's from the "Fire Zone." A fossil resting upon a rock, a root, or another object is believed to cause unusual lightning-strike appearances. These are called "Lightning Teeth," and they are stunning.

Megalodon faced an uncertain and dangerous existence like every other living thing. They had struggles with feeding injuries,[2] bite trauma, vitamin deficiency, infection, and genetics, which could all affect their teeth and their livelihood.[3]

However, scientists are applying their knowledge of tooth deformities in marine and terrestrial mammals, as well as humans, to extrapolate an understanding of this phenomenon as it applies to Megalodon.[3]

A pathological tooth is an anomaly that does not look normal. This could occur for several reasons, as mentioned above regarding a Megalodon's challenging life, including bite force tooth damage, spacing issues, and abnormal tooth replacement.[3,4]

White Megalodon Tooth. Courtesy of Mike Jacobsen.

Colorful Collection of Bull Shark Teeth.
Courtesy of Dan Case.

Battle Scars on a Great White Shark.
© iStock. Credit: Alessandro De Maddalena.

It is deformed but is still functional. These are unique. They may have wavy edges instead of straight ones, be curved in the wrong direction, look like a misshapen triangle, or have multiple tips.

When two or more crowns grow from the same tooth root, it is referred to as gemination.[3] This most common multi-tipped abnormality in teeth has been identified in a wide range of animals from bears[5] to seals,[6] dolphins,[7] mammoths,[8] and even humans.[9]

This creates the appearance of a split at the tip or can proceed down the entire length of the tooth. The tooth is still triangular-shaped and comes to a point, but technically, there are two tips instead of one.

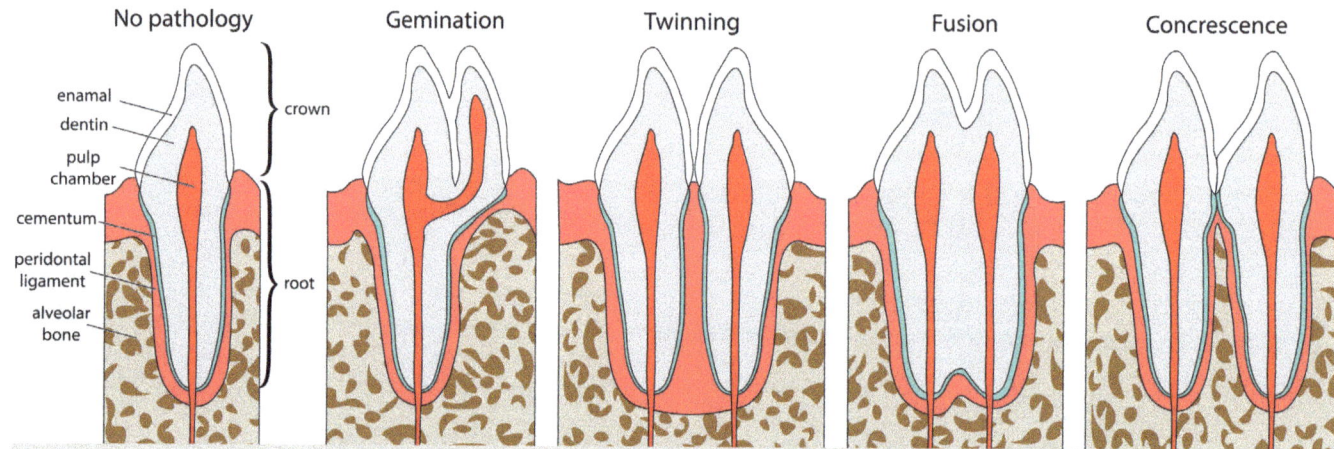

Variations in Pathological Formations of Mammalian Incisor Teeth (lingual view).
Miller, H. S. (2022). Dental pathologies in lamniform and carcharhiniform sharks with comments on the classification and homology of double tooth pathologies in vertebrates. *PeerJ, 10*, e12775. bit.ly/DentalpathologiesInSharks.
Courtesy of Harrison S. Miller.

Twinning, also known as schizodontia, is believed to occur when a tooth bud splits, allowing two independent teeth to develop in that space. They are not bonded together.[3,4] If two or more buds become connected in the early stages of tooth development, the root will produce two connected teeth, known as fusion or synodontia.[3,10]

The last oddity is concrescence. This occurs when two or more roots connect, and the teeth grow together like conjoined twins with two independent and complete teeth.[3,4,11] The classifications of tooth abnormalities as they apply to Megalodon are highly uncertain and overlap. Often, fusion and gemination cover the topic.[3,12]

A $Million Smile with an Incredible Meg Tooth.
Courtesy of Captains Blair Morrow and Michael Konecnik.
(Meg Goddess Designs and Aquanutz Scuba Diving Charters).

CHAPTER NINE

Megalodon: Dead or Alive?

Why Did Megalodon Go Extinct?

Megalodon was the tank of ocean predators, so what happened? At the end of the Miocene, approximately 7 to 5 million years ago, the world was cooling rapidly again. However, there was no direct correlation between climate change and their extinction.[1] It was more of a domino effect from numerous changes.

Approximately 3.8 to 2.4 million years ago, roughly 86% of life was centered around coastal habitats during the transition from the Pliocene to the Pleistocene.[2] This makes sense if you have ever gone snorkeling or diving. Where do you see the most sea life? By the shallows and reefs, down to around 50 feet (~15.25 m). The sunlight can penetrate the ocean, providing an ideal diving experience filled with vibrant fish and corals.

Large-scale glaciation led to dropping ocean temperatures and falling sea levels,[3,4,5] and a widening of temperatures between the low and high latitudes,[3] falling atmospheric carbon dioxide (CO_2)

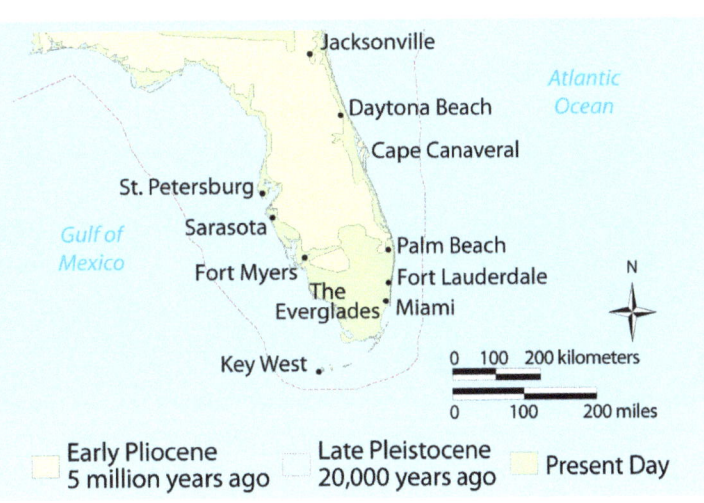

The Changing Florida Landscape. Kate Bentsen, Integration and Application Network (ian.umces.edu/media-library) from Kruczynski, W.L. and P.J. Fletcher (eds.). 2012. *Tropical Connections: South Florida's marine environment.* IAN Press, University of Maryland Center for Environmental Science, Cambridge, Maryland. 492 pp.

levels[5,6] caused by changing ocean currents and absorption in silicate rocks like basalt,[3,5] modification of oceanic nutrient distribution,[3,5] rapidly changing or loss of coastal habitats,[2] and the vulnerability associated with the energy needs of warm-blooded animals, marine life suffered the most.[2,7]

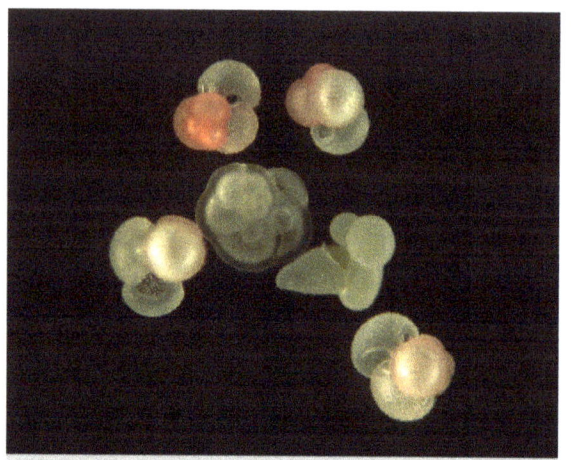

Forams from the Northern Gulf of Mexico.
Photograph by Jessica Spear (USGS).

How can we determine this information? Benthic foraminifera, commonly referred to as forams, have existed for approximately 400 million years. These tiny plankton can adapt to environmental pressures; therefore, scientists study them regularly to gain insight.

Forams are crucial for recording history as it relates to CO_2 changes,[2] the dating of marine rock layers, changes in sea level, ocean specifics like circulation, salinity, pH, climate, evolutionary modifications, ecosystems, and understanding catastrophes.[8,9]

Forams extracted carbon, calcium, and oxygen from the water, combining them like cement to form calcite. They used this calcite in conjunction with grains of sand to create an outer protective shell, known as a test. Scientists can use this calcite to help unlock mysteries in deep-sea and surface temperatures, atmospheric greenhouse gases, ice volumes, sea surface, and deep-sea temperatures.[10,11]

A Shoreline on the Potomac River. The Miocene layer, the grey marl, is exposed at the base of the cliff and into the water. I recovered several vertebrae under the water in spring as the tide and pollen kept coming in. Dr. Jay M. Lipoff.

Even today, benthic foraminifera are thoroughly researched because they have a great fossil record, and locked away inside is an incredible window into the past.[12]

Environmental changes took their toll on biodiversity, corals,[4] megafauna, and the ecosystem of that time.[3] **Between 3 and 2 Mya, almost one-third of marine life disappeared** in coastal habitats,[3] resulting in a major disruption of the food chain.

This included 55% of marine megafauna and large vertebrates, 35% of seabirds, 9% of sharks, and 43% of turtles. Mammals, like seals and dolphins, also suffered losses.[3]

Coastal habitat loss creates several issues. Previous research believed Megalodons frequented shallow waters to give their young an optimal chance of survival due to their smaller size. The latest study found that newborn Megalodons were significantly larger and could have very well survived beyond the confines of shallow waters.

Regardless, much of the prey they consumed needed these regions to survive. A changing environment leads to less habitat, nutrients, and food. As a result, the ecosystem often shifts with a loss of apex predator numbers and distribution.[13]

Research on dental calcium,[14] zinc,[15,16] and nitrogen[17] isotope levels may have unlocked crucial insights into the apex predators' domination. When Megalodon and its ancestor, *Otodus*

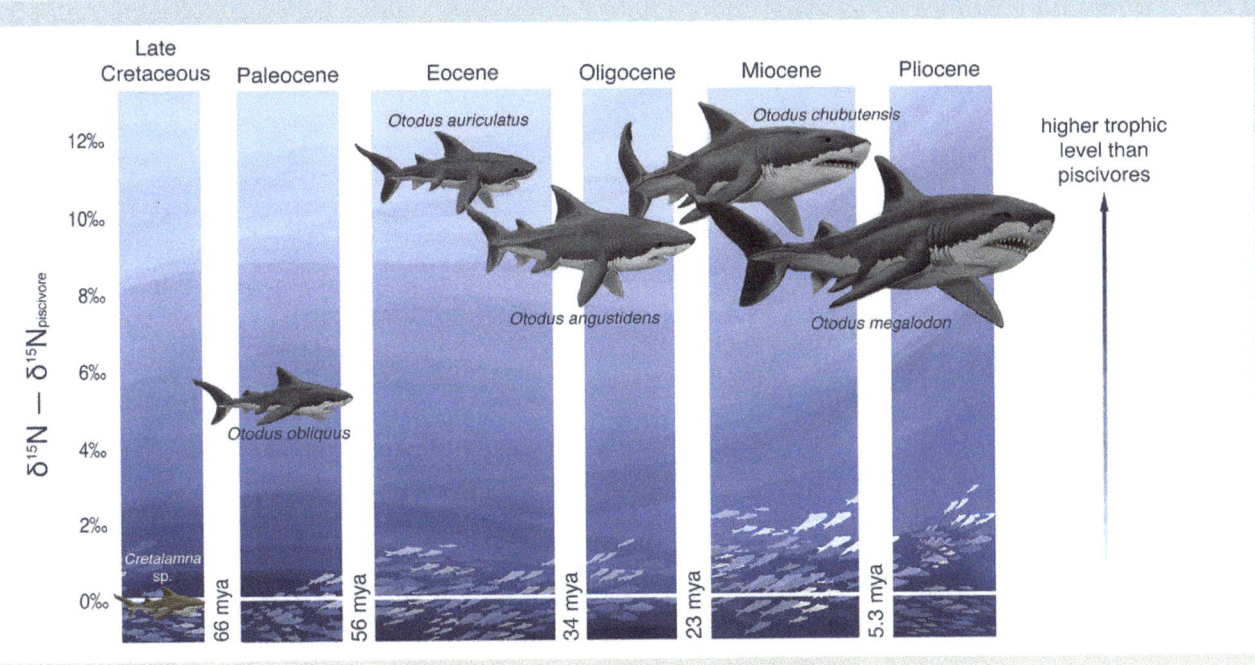

Body Size & Trophic Level of Megatooth Sharks.
© 2021. Illustration by Christina Spence Morgan. Kast, E. R. et al. (2022). Cenozoic megatooth sharks occupied extremely high trophic positions. *Science Advances, 8,* Eabl6529. bit.ly/CenozoicMegatoothSharks.

chubutensis,[18,19] were at the top of the food chain and considered the highest trophic level, their teeth's calcium and zinc levels were low, while nitrogen was high.

Five million years ago, a shift occurred, and the Megalodon's calcium and zinc levels rose while its nitrogen levels fell. A complete reversal of the isotopic levels. This meant they were no longer the big fish in a big pond. They were second fiddle to great white sharks and large predatory whales, and now faced more difficulty securing their next meal.

The great white shark and killer whales are believed to have directly competed with Megalodon for food.[20] The decline of the Megalodon's favorite prey, like filter-feeding baleen whales, had already disrupted the food chain.[2,21]

Kyptoceras during the Miocene to Pliocene. © iStock credit: CoreyFord.

Columbian Mammoths. © iStock credit: CoreyFord.

As mentioned before, due to Megalodon's high metabolic rate and the increased caloric demand needed to maintain its elevated body temperature, the disappearance of food sources was another factor that potentially contributed to its eventual demise.[22]

Struggling to find sufficient food for survival as a warm-blooded shark would lead to disaster and, ultimately, extinction. The final stages of Megalodon's demise may have occurred between 4 and 3.6 Mya.[23]

The cooler, drier air caused the lush forests to disappear, replaced by grasslands that attracted many giant herbivores and some carnivores to the region. Without ocean water being used to form glaciers and ice caps, new bridges between continents formed, bringing new life and driving the evolution of life.

Large Columbian mammoths, Woolly rhinos, and huge mastodons once covered the landscape, alongside giant sloths and tortoise, car-sized glyptodonts, herds of bison, camels deer, tapirs, and antelope, and predators such as short-faced bears, dire wolves, and saber-toothed false cats. Once again, life prevailed.

For a while, it seemed like the end of the world, but it wasn't. However, it was for the beasts of the sea. All that remains are the fearsome teeth, some vertebrae, the deep slashing predation marks on fossils, scientists' discoveries and articles, and enthusiasts who keep hope alive that one still exists in the deep blue abyss.

Ultimately, the culmination of these factors and others led to the extinction of Megalodon, the most spectacular and largest predatory shark in our planet's history.

The world can officially celebrate Megalodon's awesomeness on June 15th each year, as this has been proclaimed National Megalodon Day, so its "tail" lives on.

Could Megalodon Still Exist?

This is a million-dollar question in the minds of many fossil hunters and Megalodon enthusiasts. According to the National Oceanic and Atmospheric Administration (NOAA), we have only mapped just over 25% of the ocean floor. So, you're saying there's a chance?

Could a Megalodon, a creature of immense size and power, still be lurking in remote locations in our oceans' deepest and darkest depths, such as the uncharted depths of the Mariana Trench, which is more than 7 miles (~11.3 kilometers) deep?

Coelacanth, S. Africa, 2019.
Bruce A. S. Henderson.
(bit.ly/CoelacanthOffPumula);
Attribution 4.0 International (CC BY 4.0).

Interestingly, the prehistoric coelacanth, thought to have been extinct for 66 million years, was first caught in 1938 off the coast of South Africa. This remarkable discovery, a testament to life's resilience, was a once-in-a-lifetime occurrence. Or was it?

Remarkably, it happened again in 1952, when a fisherman hooked another one near the Comoro Islands in the Indian Ocean, off the coast of East Africa. Since 1938, over 80 more have been caught or photographed. They still exist.

Even a nautilus, once considered extinct for 500 million years, was rediscovered.[1] In 1984, an *Allonautilus scrobiculatus* was found off the coast of Ndrova Island in Papua New Guinea.[2] Then, in 1986, 2015, and most recently, *Nautilus belaunesis* was spotted at a depth in the range of 660 and 1125 feet (200 and 340 meters) in 2024.[1]

Three newly discovered and named species, *Nautilus samoaensis* from American Samoa, *Nautilus vitiensis* from Fiji, and *Nautilus vanuatuensis* from Vanuatu, can be added to the records. Maybe we haven't located them sooner because they swim to depths of up to 2,625 ft (~800 m).[2,3,4]

The deepest a great white has been logged was 4,000 feet (~1,220 m). Therefore, most scientists believe that no shark has been recorded as deep as 10,000 feet (~3,048 m) due to the absence of light, water temperatures of 34 to 39 °F (~1 to 4 °C), and a lack of large quantities of food.[5]

Nautilus.
Photo by Michelle Johnston/NOAA.

Based on that information, it is highly unlikely that a Megalodon is breaking records in the Mariana Trench in the western Pacific Ocean, unseen for millions of years, patrolling the darkness, and residing at an estimated depth of 7 miles or 36,070 feet (~11,278 m).

Nevertheless, a 2023 research article reported the discovery of the tip of an *Otodus megalodon* tooth on the undisturbed Pacific Ocean floor at a depth of over 1.9 miles or 10,000 feet (~3.06 km).[6]

Other Megalodon teeth and numerous whale ear bones have been found at depths ranging from 0.28 to 3.46 miles (~0.45 to 5.57 km). So, perhaps Megalodon once migrated across these open waters or utilized them for mating or chasing meals.[6]

Feeding and the regular tooth replacement of sharks would easily explain why teeth are found a long way from any coastline, typically at a distance greater than 560 miles (~900 km) from dry land.[6]

Some believe they have encountered Megalodon and saw the incredible beast that still haunts the darkness.[7,8] It would be cool to see it, but like the Loch Ness monster and Bigfoot, we should have had solid evidence of its existence by now.

Can we go anywhere without someone having a camera phone? As with any extraordinary claim, it is essential to evaluate the evidence critically. At this point, you are safe to return to the water.

There Are Two Sides to Every Story. A Meg tooth with a clean front and a barnacle-covered back. Megalodon may be gone, but it will never be forgotten. Courtesy of Blair Morrow and Michael Konecnik. (Meg Goddess Designs, Aquanutz Scuba Diving Charters).

Nice Finds by Kristina Palumbo (Coastal Collector by Nautical Necklaces). Courtesy of Kristina Palumbo.

CHAPTER TEN

Fossil Hunting

The Basic Strategies of Fossil Hunting

Grab your geology hammer, chisel, shovel, dive gear, sunglasses, snacks, and the family. Have an adventure, be patient, and please be safe. Like any other sport or activity, it takes practice, repetition, instruction and even videos to make you into a better fossil finder.

You may also want to add safety glasses, bug spray, several pairs of gloves, a collection bag, a towel for sweating and drying off, a tent or umbrella for shade, a folding chair, and sunscreen to your list.

Before you start destroying the Earth, look into the laws of your state and see if you need to fill out any special licensing forms to go fossil hunting legally.

For instance, some states limit the removal of fossils to those below the waterline, which varies depending on the tide levels. Florida Statute 1004.57 requires obtaining a permit for private collecting, which isn't too expensive; however, if you start recovering archaeological finds, it is illegal.

Finding a Fossil on the Shore.
Courtesy of Cooper N. Stephens.

Never start digging a deep hole along the riverbanks or on the border of someone's private property; always ask for permission. **Without being cautious or considerate, you could cause major erosion or collapse**, potentially harming those in your fossil-hunting party.

Potomac River, Virginia (left). **Calvert Cliffs, Chesapeake Bay, MD** (center & right).
My two young fossil explorers by the gray marl, and my father-in-law, Ray, on his airboat. Dr. Jay M. Lipoff.

Ask questions and talk to experts and locals online or in their shops. Hire a guide or business to show you how to be successful when you explore fossil beds on land or in the water. South Carolina, Maryland, and Florida's West Coast are great locations to visit.

On a shoreline, look where the gravel is pushed up along the beach from the tide. If it's low tide, more material will be visible.

I recommend picking a row of gravel that is exposed to browse while the water is out. Move away from the water and then back in a zigzag pattern. Cover the area before the tide returns. You can cover everything and take as much time as you need.

Aurora, Fossil Museum, N.C.
Ray Harding screening fossils.
Dr. Jay M. Lipoff.

If you are familiar with your area, it may produce a specific type of fossil, such as black fossils, which can help you identify a potential find by recognizing the color.

Another technique is to use a screening scoop. Drag the scoop to fill it, then shake it carefully in the water to clean up your finds. The sand and smaller items will fall through the holes, leaving the shells, pebbles, and your fossils behind.

Start to look for triangle shapes or shimmering from the glossy enamel, and you may be on to teeth. This can also help you achieve success. Fossils are usually dense and have a glassy timbre when tapped, which is distinctly different from that of rock. Wait til you see how many leaves, shells, or buried rocks make you think you have found a beauty. Deception. Plain and simple.

The shorelines are fun, and North Carolina, South Carolina, Florida, and Maryland have miles of them, concealing wonderful fossils. You can find them in the shallow waters, along the beach, and in the gray marl of the cliffs. Remember, however, that no one is supposed to dig in the cliffs. It damages them and can lead to a potentially hazardous collapse.

There are also locations for land-based fossil hunting where you can dig with both feet firmly planted on the ground, like Aurora Fossil Museum in Aurora, N. C., Palmetto Fossil Excursions (PFE) in Summerville, S. C., the Bone Valley Experience in Wauchula, Florida, Bone Valley Fossil Farms in Bowling Green, Florida, and the Calvert Cliffs Region in Calvert County, Maryland. Each has produced some beautiful fossils.

The little museum in Aurora, N. C., is great. This region once allowed fossil hunters to explore the phosphate mines and retrieve glorious megalodon teeth.

After a while, they shut it down, but the quarry continues to bring screened material for families to sift through and find many smaller fossils.

My Boys Digging and Screening for Fossils at Palmetto Fossil Excursions (P.F.E.).
Dr. Jay M. Lipoff.

An umbrella or a tent is wise to block the sun's heat for the other locations mentioned. The sun can be relentless, and the area will get muddy, so bring clothes to change into before leaving and a trash bag for the muddy gear. Lots of water and snacks are a must, as is bug spray.

PFE has a screening station where you can stand and spray water on the material brought to your station. This offers a nice break from digging in wet clay; your knees may thank you. They provide shovels, but pack plenty of gloves to protect your hands.

Bone Valley Fossil Farm.
Bowling Green, Florida. Courtesy of BVFF.

Bone Valley brings you a bucket loader filled with dirt from a fossil-yielding layer, and you spray water to reveal fossils. The quicker you finish your pile, the sooner the next load of dirt will be brought. They provide shovels and portable screening stations.

There is also the adventure of diving. One prime location is the waters off the west coast of Florida, where a bounty of fossils and several knowledgeable operations await, taking you to the hot spots they have researched.

The water is seasonably warm, which will determine the thickness of your dive suit or dive skin. Visibility ranges from 3 to 20 feet (~0.9 to 6.1 m), occasionally resembling egg-drop soup or what many refer to as "whale snot." It can be as murky as a dust storm.

Teeth Hidden in Plain Sight. Megalodon Teeth (left & center). **Snaggletooth** (right).
Gulf Waters off Venice, Florida. Courtesy of Kristina Palumbo (Nautical Necklaces & Aquanutz Scuba Diving Charters).

Due to challenging visibility and the absence of reefs, it's easy to feel lost or off track. If this happens, you can always surface while following standard diving safety protocols, as you are less than 30 ft. (~9.1 m) below the surface in most cases.

For instance, if you did not follow the pre-dive instructions to the letter, follow your bubbles up to the surface with your hand extended upward, relocate the boat and shoreline to get your bearings, swim back to the bottom, and continue searching for teeth. Not all boaters honor or recognize a dive flag when it is displayed. Some come much closer than the law allows.

Typically, you focus on searching the black gravel rather than shell beds. Some folks cruise quickly to cover as much ground as possible, while others carefully move their bodies from side to side to cover a larger swath completely.

A Monster from a Carolina Ledge.
A 6.5-inch (~16.5 cm) long by 5-inch (~12.7 cm) wide Megalodon tooth. Courtesy of Ryan Picou.

Some folks like to fan the bottom. This technique involves moving your hands to clear a space, but a lack of current will create a slow-dispersing dust cloud for you and anyone else.

For deep-sea diving and many large teeth, nothing compares to a trip where you boat two hours offshore from North Carolina and then plunge 80 to 100 feet (~24.4 to 30.5 m) to the world-famous meg-lined ledge.

Tether your line to the anchor and quickly explore the ledge. Keep track of your bottom time, which is limited by your depth, and reserve some air for safety stops.

The good news is that diving operations don't share their locations, so the number of people visiting is limited to their excursions. This means big teeth and handfuls of them. Some divers use rebreathers to stay underwater longer and find more teeth.

Many states, including Maryland, Virginia, North Carolina, South Carolina, Georgia, and Florida, offer shallow creeks and rivers loaded with fossils if you want to search and explore shallow water. Some are along the shores, but more lie beneath the dark water.

Blackwater fossils are spectacular. Some people travel by foot, boat, jet ski, or kayak to reach their destination. Once you arrive, you can dive, snorkel, wade, or walk with a floating screening apparatus.

They scoop up a shovel full of potential gravel-yielding material and sift it in the water to expose their rocks, shells, and, hopefully, fossils.

Underwater, there can be some intense currents, visibility is less than a few feet (~1 m), and you need to know the waters and behavior of the resident alligators.

It is their world, and they rarely bother humans, but you want to ensure that when you grab that tooth and yank it out of the clay, it is not an alligator. It's their world. Get a guide or research their habits.

Creek Hunt with P.F.E. in South Carolina.
Closely examining the material on the screen.
Dr. Jay M. Lipoff.

Black Water Creek Find.
Courtesy of Michael Tyler Staab
(I Hunt Dead Things).

Black Water Creek Hunt.
Courtesy of Clay Cook.

Preparation Tips

If you find fossils with barnacles, coral, or other types of encrustation covering them, you could leave them alone. These are so beautiful, I would. Otherwise, there are methods to remove them without damaging the fossil.

In general, **water discoveries require a good soak to remove salt, odors, and impurities, which can help stabilize the fossil.** Usually, a few days are sufficient. Some people recommend placing their larger fossils in the toilet tank, not the bowl, so they receive a gentle freshwater wash with each flush.

Growth On Meg Tooth.
Courtesy of Blair Morrow.

Coral Growth on Fossil.
Dr. Jay M. Lipoff.

Another strategy involves using a vinegar and water solution to help dissolve calcium carbonate. Sometimes, the growth is perfectly placed and should be left for aesthetic reasons. Ask for opinions if you are unsure. It can make a fossil extra unique.

Depending on the growth level, I use a vinegar concentration between 25% and 50%. This ratio varies according to your goals, so it's essential to monitor progress closely.

You can gently sandblast with a microabrasive, like baking soda, or use a wire brush, a toothbrush, a dental pick, or a bench grinder with a wire brush wheel. However, test it first to be safe.

If your fossil is damaged but still a valuable find, you can find collectors who know how to repair it, or they will know someone who can restore it to its original condition or nearly perfect condition.

Join a fossil or Megalodon social media group for help locating, identifying, cleaning, finding guided excursions, understanding value before purchasing, learning to recognize natural fossils versus those with repairs that reduce their overall value, and for fossil hunter friendships.

Also, search fossil-hunting blogs, YouTube videos, and websites for similar advice and tips. This is a great way to get started. You will make amazing memories and friendships, and you will discover incredible treasures.

Bench Grinder. Courtesy of Blair Morrow. (Meg Goddess Designs).

After speaking with a young volunteer, Braum Tokarski, I was permitted to explore the curator's room during my recent visit to the Calvert Marine Museum in Solomons Island, Maryland.

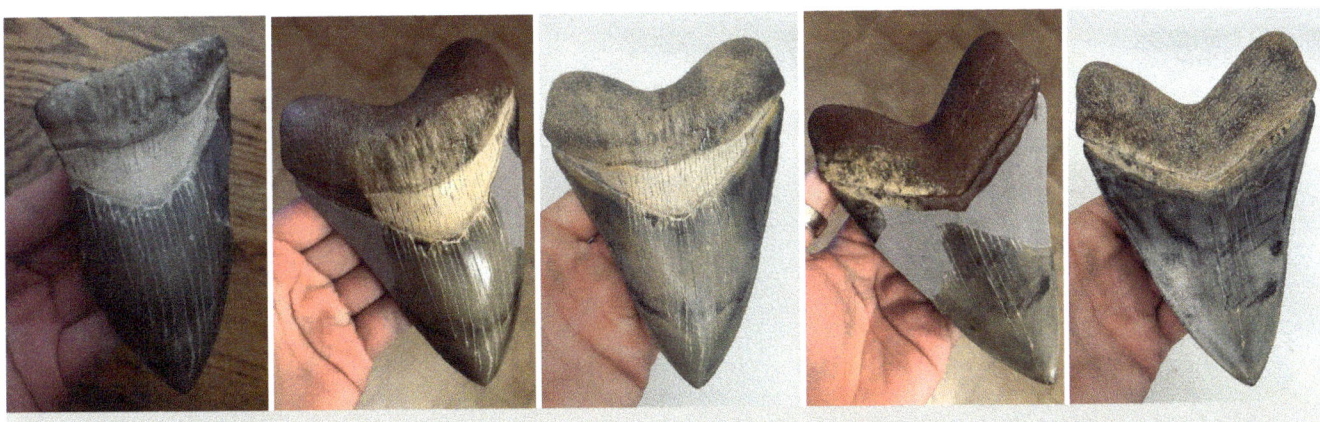

Repaired Tooth. (On the far left is the original, Pic. #1. Check out the finished product, Pic. #5.) Images 2 and 4 show the repairs in progress, and 3 and 5 are the final product. An untrained eye would pay too much for this. Always check your sources and ask for advice. Courtesy of Blair Morrow (Meg Goddess Designs, Aquanutz).

He spoke to me about the findings they were working on, the ones on the shelves, and I'm sure I acted like an overexcited child on Christmas morning. He was cleaning a fossilized dolphin skull with a wooden dowel, that had been awaiting attention for decades beneath densely packed sand, not the one pictured below.

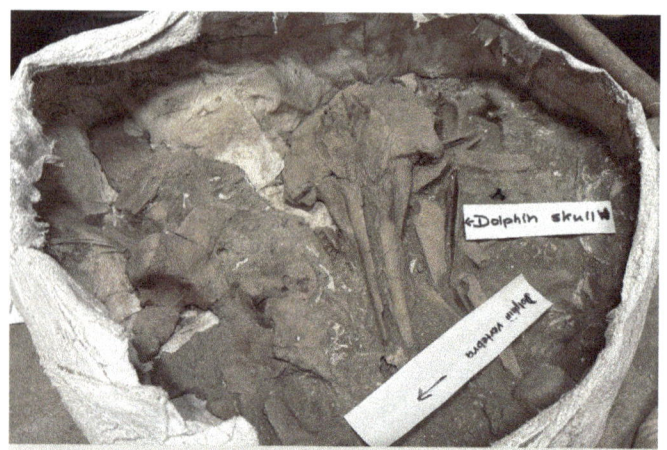

Dolphin Skull in Matrix in Plaster Cast.
Permission by Dr. Stephen J. Godfrey. Dr. Jay M. Lipoff.

Vertebrae in Matrix in a Resin Cast.
Permission by Dr. Stephen J. Godfrey. Dr. Jay M. Lipoff.

These are two excellent examples of professional practices used to safely transport a dolphin's skull and vertebrae (left), where they were carefully shielded by plaster. Hardening resin is sometimes used, as seen in the vertebrae image on the right.

The final piece of both specimens will be revealed after extensive work with tools such as dental picks, an air scribe, chemical etching solutions, and a combination of techniques and skills that require considerable time and patience. Also, perhaps not being interrupted by an enthusiastic writer and fossil explorer.

The next 200 images will captivate you with the excitement of discovering your own fossils and expanding your collection. Here are a few pictures of fossils collected from areas along the East Coast.

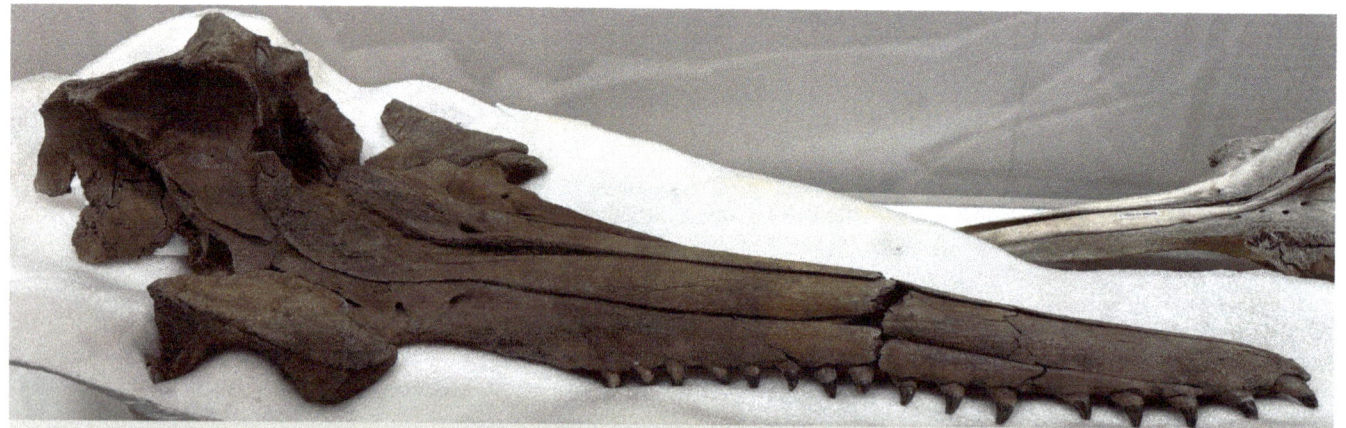

A Dolphin Skull Impeccably Cleaned. On display at the Calvert Marine Museum. Dr. Jay M. Lipoff.

Nice Finds Diving Venice, Florida. Dr. Jay M. Lipoff.

You Might Find a Unicorn, So Bring the Calipers.
Courtesy of Blair Morrow. (Meg Goddess Designs, Aquanutz).

**From a Creek Hunt in South Carolina
with Palmetto Fossil Excursions**.
Dr. Jay M. Lipoff.

A Bounty of Fossils. Dr. Jay M. Lipoff.
After several outings along the Potomac River in Maryland, I collected and soaked these fossils in water for 24 hours. Next, I set the fossils on the picnic table in the sun to dry, making the blue marl clay easier to remove with a dental tool. This is not a typical haul. There are lots of whale vertebrae and rib bones on the table. In my hand is an atlas vertebra.

A Shark Vertebra and Lightning Teeth Waiting to be Found. Courtesy of Erin Osborne. (Charleston Center for Paleontology).

Large Vertebra.
Courtesy of Aquanutz Scuba Diving Charters.

Columbian Mammoth Skeleton.
(Florida Museum of Natural History). Dr. Jay M. Lipoff.

Mastodon Skeleton.
(Florida Museum of Natural History). Dr. Jay M. Lipoff.

 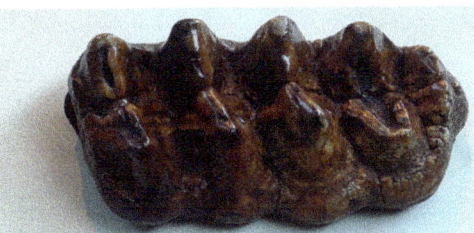

Comparison Between the Surface of a Mammoth Tooth (left) **and a Mastodon Tooth** (right).
On display at the Florida Museum of Natural History, Gainesville, Florida. Dr. Jay M. Lipoff.

Fossil Identification Charts

Identifying fossils can be challenging. The Internet and certain Facebook groups can be quite helpful. Some people compare their findings to drawings, while others succeed by examining images. Here are a couple of samples for your review.

The more you examine them and study the differences, the more accurately you will be able to identify them. If you can't figure it out, there are Facebook pages where you can post for assistance. Many museums also allow you to bring in fossils for identification.

There are ID Charts with drawings and photos contributed by extremely knowledgeable individuals online. This is not an all-inclusive list of fossils. Many more species exist, but are rare to find. On some pages, you will need to turn the book sideways, but the sheets are so fantastic, I wanted them to be large enough to read.

If you can't remember all the identifying features of the fossils, take a picture and make it a favorite. This way, you can refer back to it as needed. Good luck.

The remaining pages contain over 200 images of real fossils, which will help you understand and identify them more easily. Or, if you're like me, you wonder how they got so lucky and why I can't seem to find these incredible fossils when I'm looking?

Common Fossil Haunts
IN FLORIDA

For more information check the web

Amelia Island

Gainesville Museum of Natural history

Yankeetown

Honeymoon Island

Venice Beach

Peace River

Any shell pit along Florida's coastline is likely to hold fossils

Coastline during Miocene

Coastline during Plio-Pleistocene

Florida first began to rise from the Ocean 30+ MYA

Common Fossil Haunts in Florida. Courtesy of Russel Brown.

Fossil Sharks of Florida

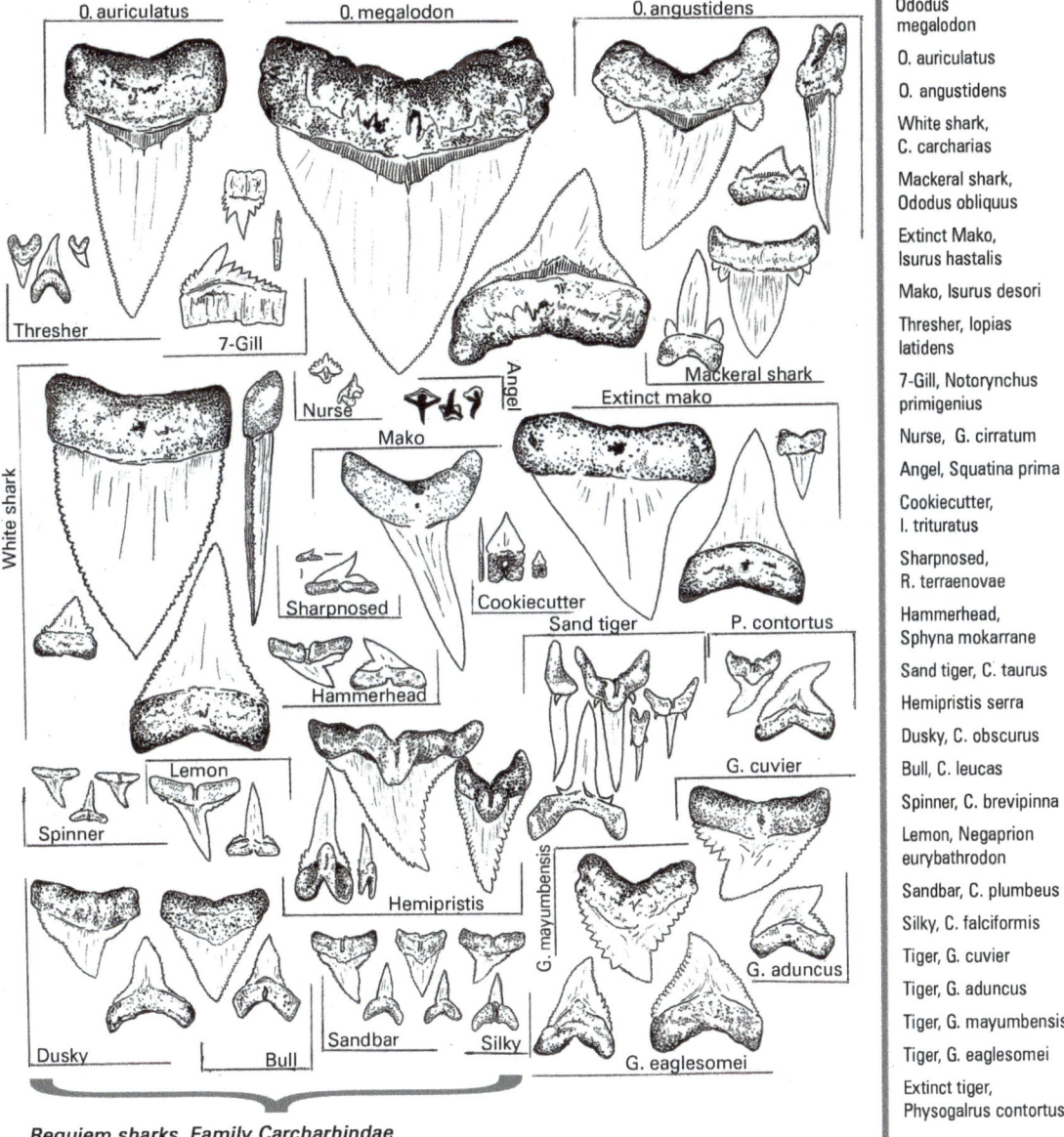

Species list (right column):

Ododus megalodon

O. auriculatus

O. angustidens

White shark, C. carcharias

Mackeral shark, Ododus obliquus

Extinct Mako, Isurus hastalis

Mako, Isurus desori

Thresher, Iopias latidens

7-Gill, Notorynchus primigenius

Nurse, G. cirratum

Angel, Squatina prima

Cookiecutter, I. trituratus

Sharpnosed, R. terraenovae

Hammerhead, Sphyna mokarrane

Sand tiger, C. taurus

Hemipristis serra

Dusky, C. obscurus

Bull, C. leucas

Spinner, C. brevipinna

Lemon, Negaprion eurybathrodon

Sandbar, C. plumbeus

Silky, C. falciformis

Tiger, G. cuvier

Tiger, G. aduncus

Tiger, G. mayumbensis

Tiger, G. eaglesomei

Extinct tiger, Physogalrus contortus

Labels within figure: O. auriculatus, O. megalodon, O. angustidens, Thresher, 7-Gill, White shark, Nurse, Angel, Mako, Mackeral shark, Extinct mako, Sharpnosed, Cookiecutter, Sand tiger, P. contortus, Hammerhead, G. cuvier, Lemon, Spinner, Hemipristis, G. mayumbensis, Dusky, Sandbar, Silky, Bull, G. aduncus, G. eaglesomei

Requiem sharks, Family Carcharhindae
This family of sharks has 60 species and is one of the largest family of sharks. Most have triangular, blade-like upper teeth with small serrations. Lower teeth are spike-like with broad bases. Teeth are very similar among the different species, making it difficult to identify the species in this family group.

Fossil Sharks of Florida. Courtesy of Russel Brown.

Florida Fossil Hunters ID Sheet

Mammals

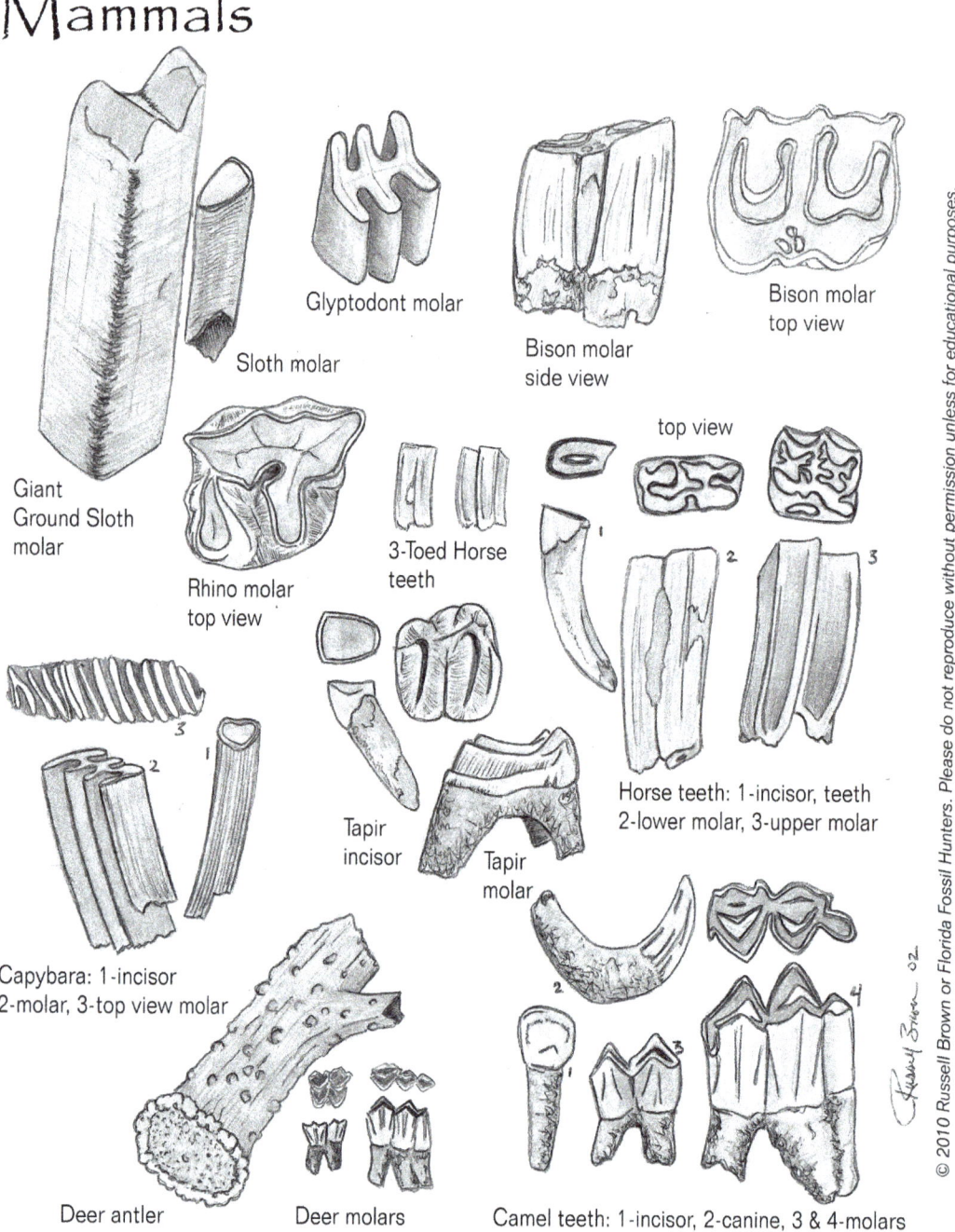

Glyptodont molar

Sloth molar

Bison molar side view

Bison molar top view

Giant Ground Sloth molar

Rhino molar top view

3-Toed Horse teeth

top view

Horse teeth: 1-incisor, teeth 2-lower molar, 3-upper molar

Tapir incisor

Tapir molar

Capybara: 1-incisor 2-molar, 3-top view molar

Deer antler

Deer molars

Camel teeth: 1-incisor, 2-canine, 3 & 4-molars

Florida Fossil Hunters ID Sheet. Courtesy of Russel Brown.

Common Shark Fossils from the Miocene and Pliocene of North and South Carolina

For Detailed Information Go to:

www.fossilguy.com

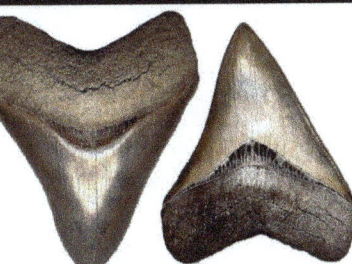

Otodus megalodon
(Megatooth Shark)

Otodus auriculatus
(Megatooth Shark)

Carcharodon carcharias
(Great White Shark)

Carcharodon plicatilis
(Giant White Shark)

Carcharodon hastalis
(Narrow White Shark)

Isurus oxyrhinchus
(Mako Shark)

Hemipristis serra
(Snaggletooth Shark)

Notorynchus cepedianus
(Sixgill Cow Shark)

Hexanchus griseus
(Sevengill Cow Shark)

Galeocerdo cuvier
(Tiger Shark)

Galeocerdo aduncus
(Tiger Shark)

Physogaleus contortus
(Tiger-like Shark)

Sphyrna sp.
(Hammerhead Shark)

Carcharhinus sp.
(Requiem Shark)

Carcharias sp.
(Sand Tiger Shark)

Negaprion sp.
(Lemon Shark)

Shark Vertebra

ID Sheets of North and South Carolina. Courtesy of Jayson Kowinsky (FossilGuy.com).

COMMON VERTEBRATE FOSSILS FROM THE MIOCENE OF MARYLAND AND VIRGINIA

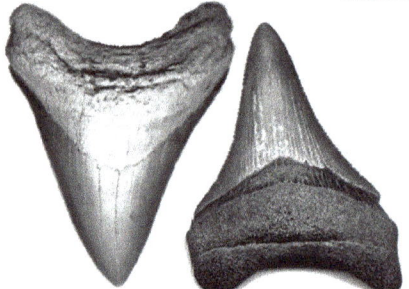

Shark Vertebra

Otodus megalodon
Extinct Megatooth Shark

Otodus auriculatus
Extinct Megatooth Shark

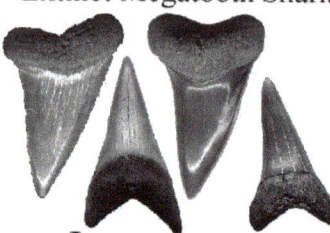

Galeocerdo aduncus
Extinct Tiger Shark

Isurus sp.
Mako Shark

Hemipristis serra
Snaggletooth Shark

Galeocerdo contortus
Extinct Tiger Shark

Squatina subserrata
Angel Shark

Alopias latidens
Thresher Shark

Notorhynchus cepedianus
Cow Shark

Carcharias sp.
Sand Tiger Shark

Carcharhinus sp.
Requiem Shark

Sphyrna sp.
Hammerhead Shark

Negaprion sp.
Lemon Shark

Porpoise teeth

Ray teeth/pavements

WWW.FOSSILGUY.COM

ID Sheets of Maryland and Virginia. Courtesy of Jayson Kowinsky (FossilGuy.com).

SCFossils.com Summerville Shark Tooth Guide

Otodus megalodon

Otodus angustidens

Carcharodon hastalis
(Lesser White)

Carcharodon carcharias
(Great White)

Galeocerdo aduncus
(Extinct Tiger)

Physogaleus contortus
(Extinct Tiger)

Galeocerdo cuvier
(Modern Tiger)

Hemipristis serra
(Snaggletooth)

Isurus desori
(Shortfin Mako)

Isurus retroflexus
(Longfin Mako)

Carcharoides catticus

Carcharias sp.
(Sand Tiger)

Alopias Grandis
(Giant Thresher)

Alopias vulpinus
(Cusped Thresher)

Parotodus sp.
("Cusped Benedenii")

Parotodus Benedenii
(False Mako)

Notorhynchus sp.
(Cow Shark)

Sphyrna sp.
(Hammerhead)

Squatina sp.
(Angel shark)

Gingliostymidae
(Nurse Shark)

Carcharinus leucas
(Bull Shark)

Summerville Shark Tooth Guide. Courtesy of Jared Shuler (www.SCFossils.com).

Common Fossilized Shark Teeth of Florida

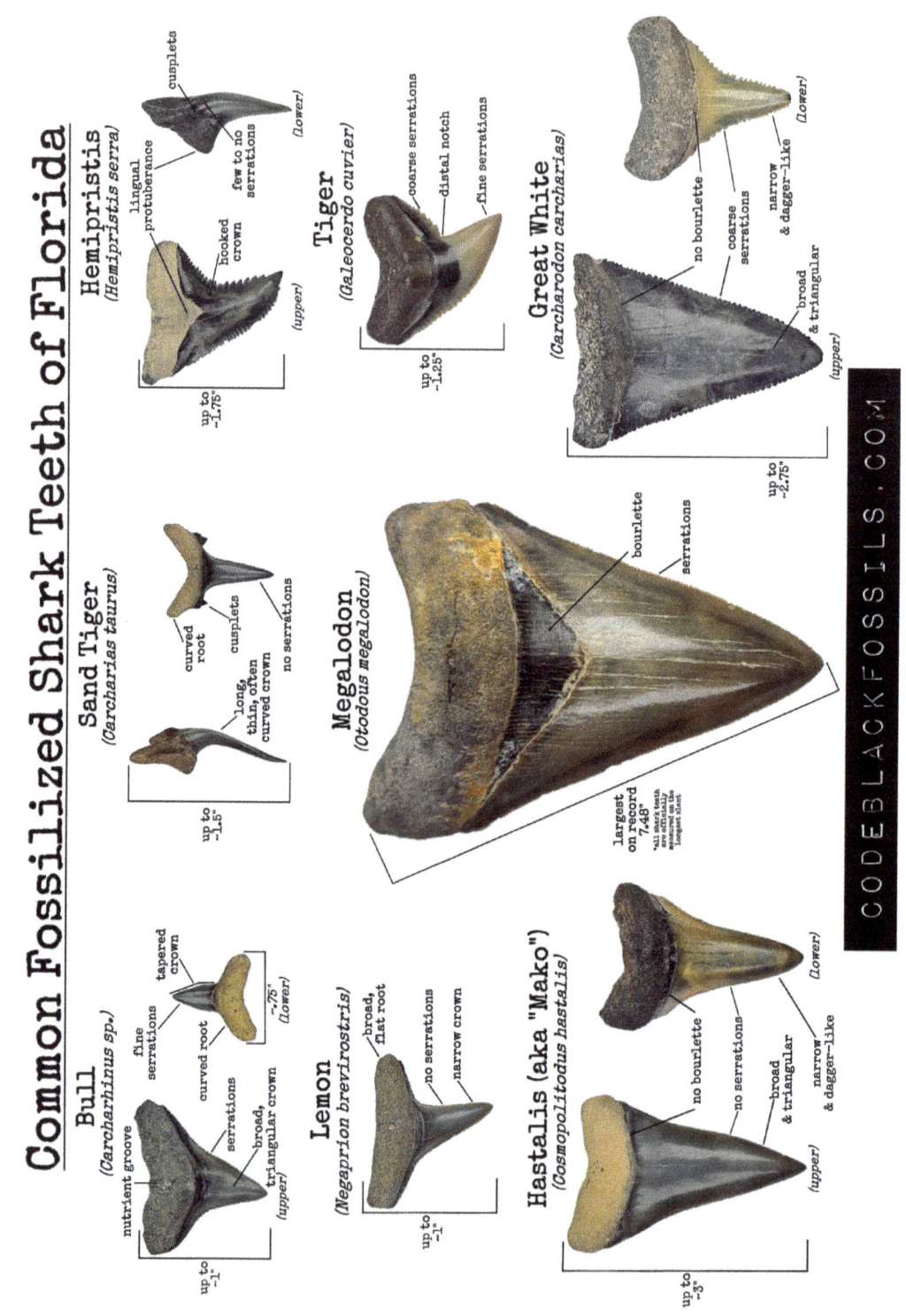

Hemipristis
(Hemipristis serra)

cusplets

lingual protuberance

few to no serrations

hooked crown

(lower)

(upper)

up to ~1.75"

Tiger
(Galeocerdo cuvier)

coarse serrations

distal notch

fine serrations

up to ~1.25"

Great White
(Carcharodon carcharias)

no bourlette

coarse serrations

narrow & dagger-like

(lower)

broad & triangular

(upper)

up to ~2.75"

Sand Tiger
(Carcharias taurus)

curved root

cusplets

long, thin, often curved crown

no serrations

up to ~1.5"

Megalodon
(Otodous megalodon)

bourlette

serrations

largest on record 7.48"
*all sizes measured on the longest slant

Bull
(Carcharhinus sp.)

tapered crown

fine serrations

curved root

~.75" (Lower)

nutrient groove

serrations

broad, triangular crown

(upper)

up to ~1"

Lemon
(Negaprion brevirostris)

broad, flat root

no serrations

narrow crown

up to ~1"

Hastalis (aka "Mako")
(Cosmopolitodus hastalis)

(Lower)

no bourlette

no serrations

narrow & dagger-like

broad & triangular

(upper)

up to ~3"

CODEBLACKFOSSILS.COM

ID Sheets of Maryland and Virginia. Courtesy of Jayson Kowinsky (FossilGuy.com).

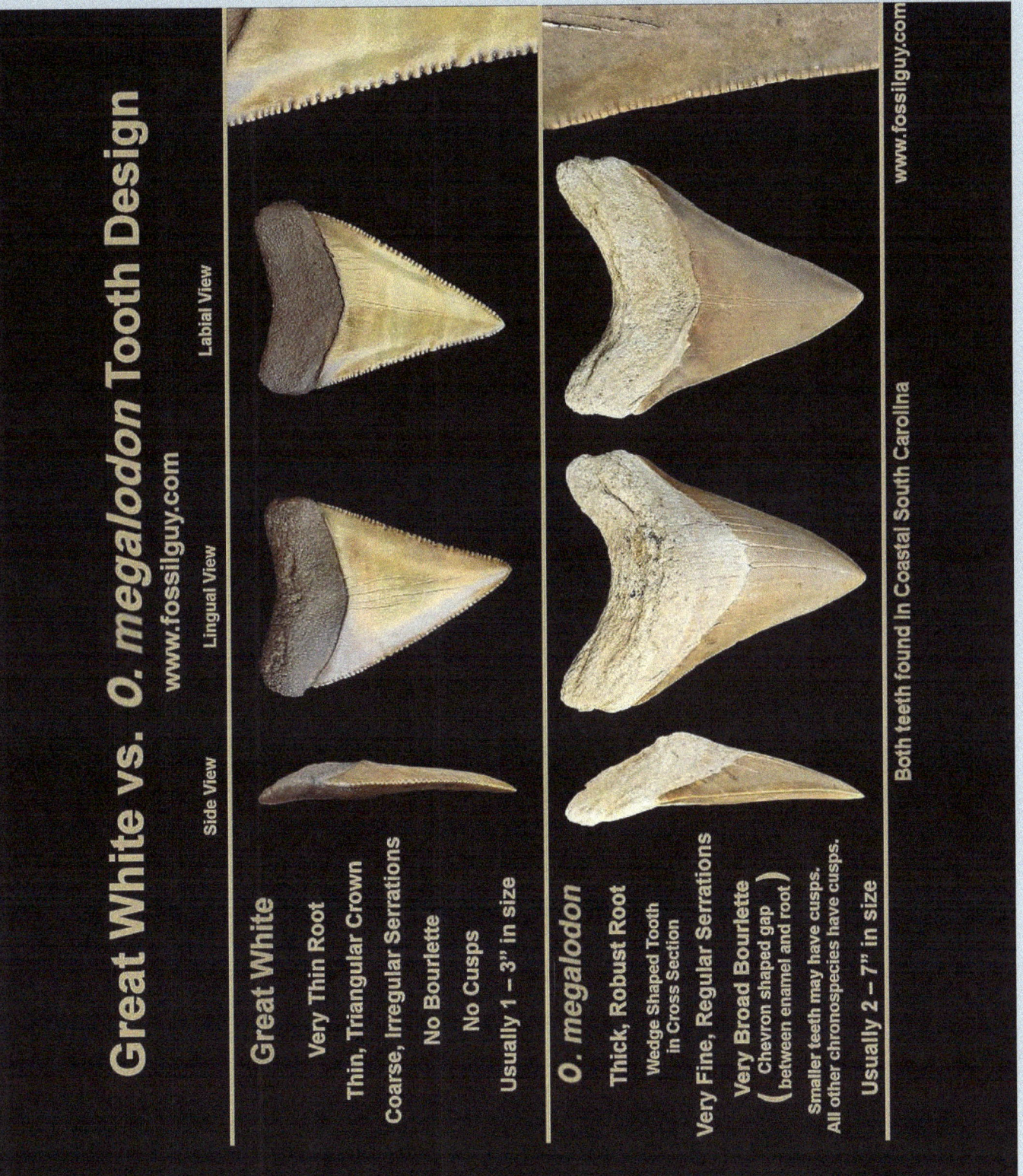

Great White vs. *O. megalodon* Tooth Design

www.fossilguy.com

Labial View **Lingual View** **Side View**

Great White

Very Thin Root

Thin, Triangular Crown

Coarse, Irregular Serrations

No Bourlette

No Cusps

Usually 1 – 3" in size

O. megalodon

Thick, Robust Root

Wedge Shaped Tooth
in Cross Section

Very Fine, Regular Serrations

Very Broad Bourlette
(Chevron shaped gap)
(between enamel and root)

Smaller teeth may have cusps.
All other chronospecies have cusps.

Usually 2 – 7" in size

Both teeth found in Coastal South Carolina

www.fossilguy.com

Megalodon vs Great White Differences Sheet. Courtesy of Jayson Kowinsky (FossilGuy.com).

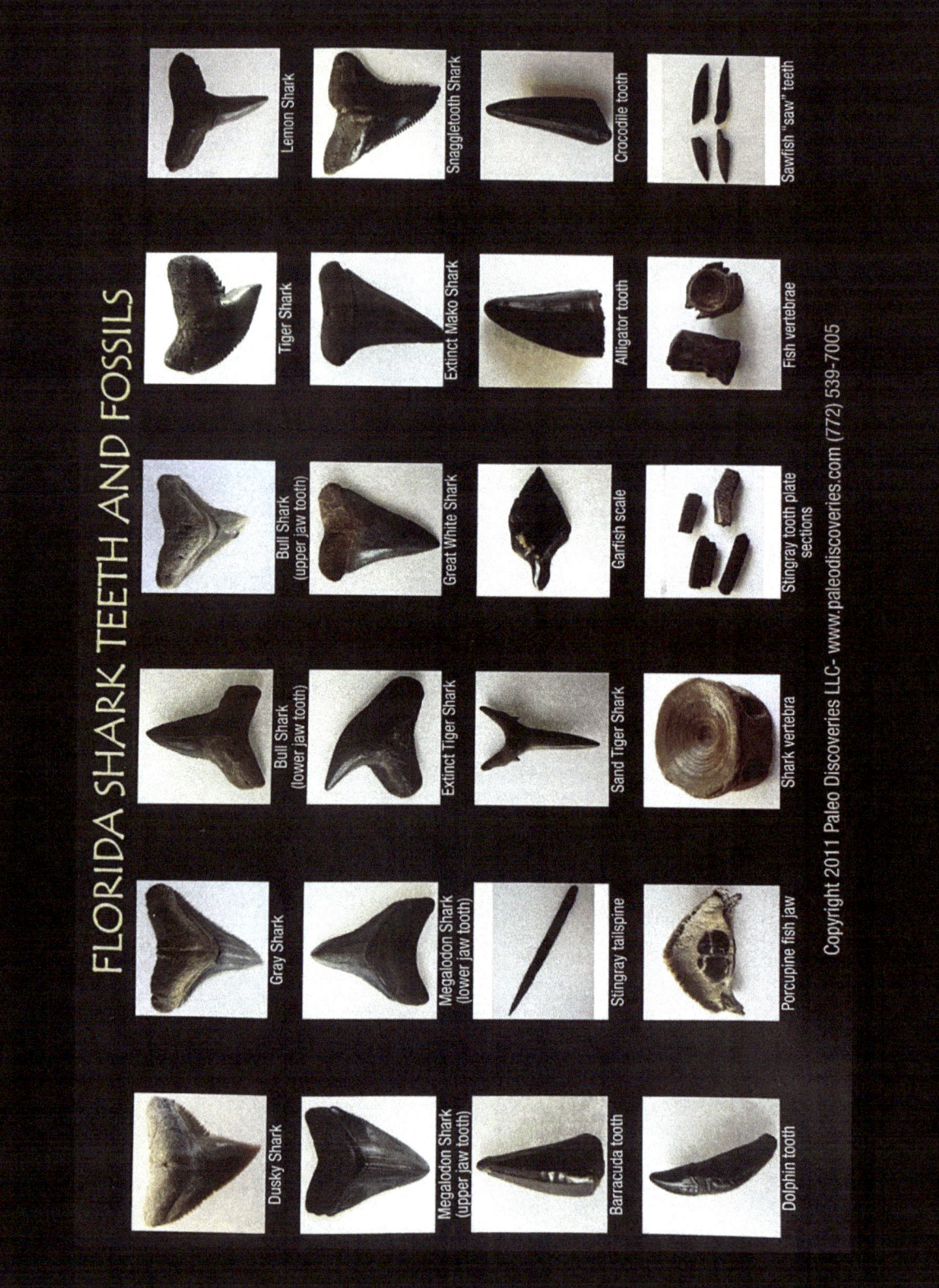

FLORIDA SHARK TEETH AND FOSSILS

Lemon Shark

Snaggletooth Shark

Crocodile tooth

Sawfish "saw" teeth

Tiger Shark

Extinct Mako Shark

Alligator tooth

Fish vertebrae

Bull Shark (upper jaw tooth)

Great White Shark

Garfish scale

Stingray tooth plate sections

Bull Shark (lower jaw tooth)

Extinct Tiger Shark

Sand Tiger Shark

Shark vertebra

Gray Shark

Megalodon Shark (lower jaw tooth)

Stingray tailspine

Porcupine fish jaw

Dusky Shark

Megalodon Shark (upper jaw tooth)

Barracuda tooth

Dolphin tooth

Copyright 2011 Paleo Discoveries LLC- www.paleodiscoveries.com (772) 539-7005

Florida Shark Teeth and Fossils ID Card. Courtesy of Frank Mazza (Paleo Discoveries).

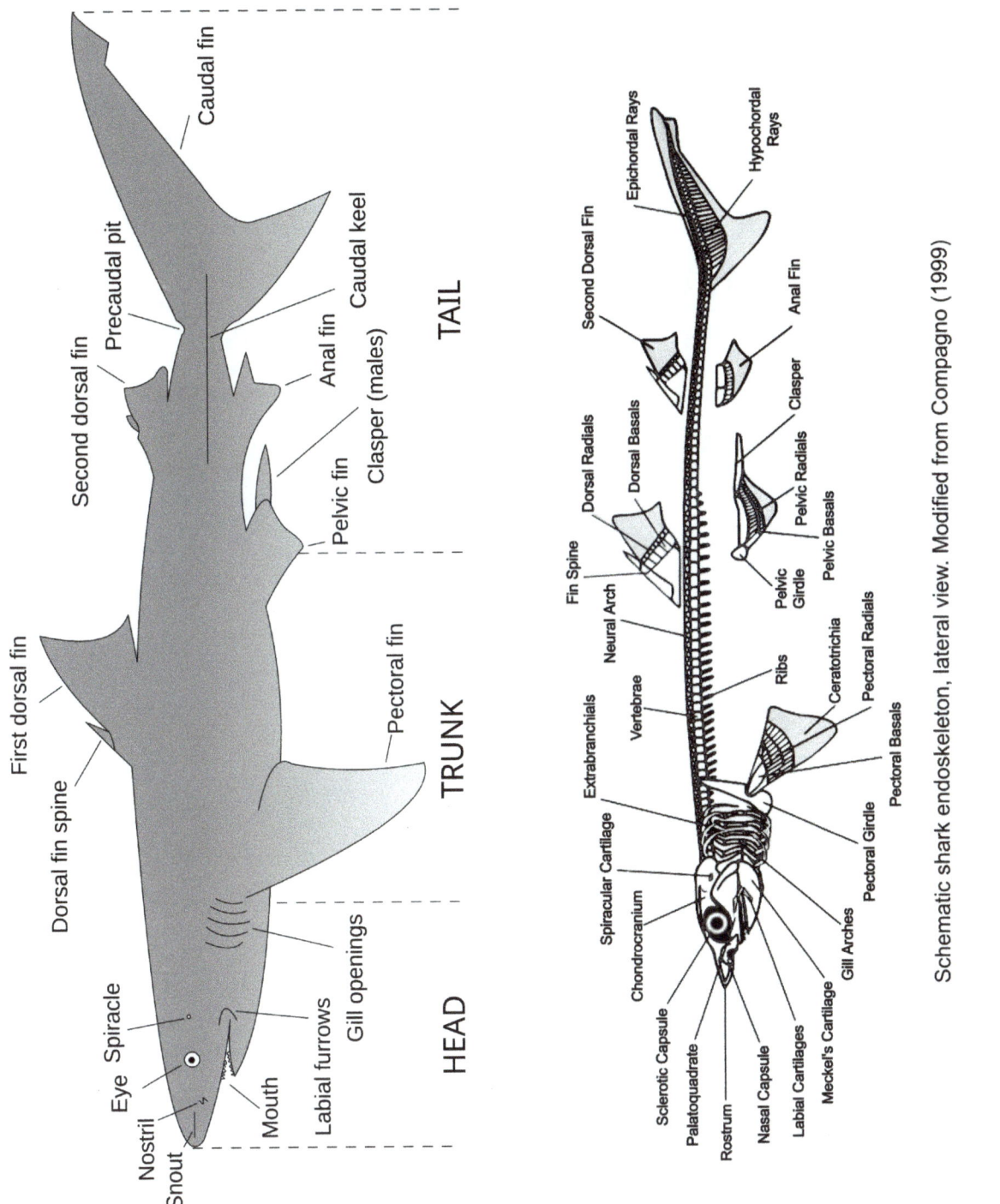

Shark Endoskeleton. Enault, S.; Auclair, C.; Adnet, S. & Debiais-Thibaud, M. (2016). A complete protocol for the preparation of chondrichthyan skeletal specimens. *Journal of Applied Ichthyology. 32(3)*: n/a-n/a. bit.ly/PrepChondrichthyanSkeletal. Courtesy of Melanie Debiais-Thibaud (Université de Montpellier). Redrawn by Dr. Jay M. Lipoff.

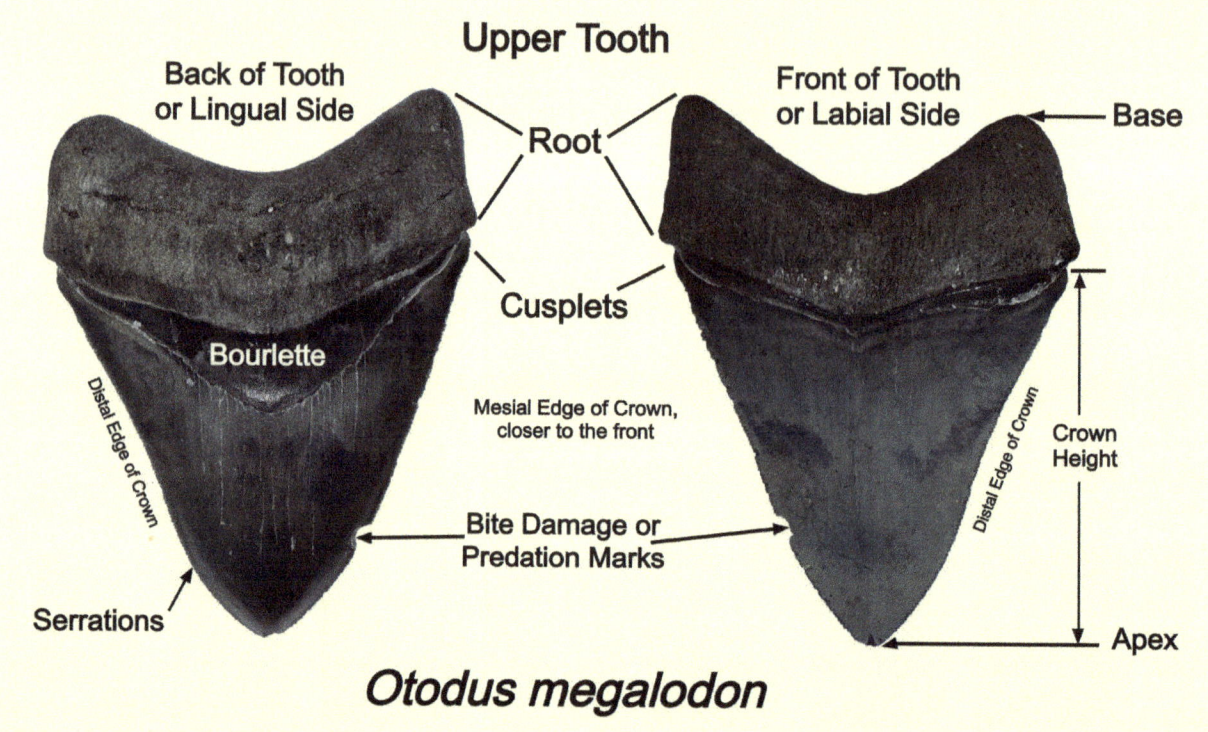

Upper Tooth

Back of Tooth or Lingual Side

Root

Front of Tooth or Labial Side

Base

Cusplets

Bourlette

Distal Edge of Crown

Mesial Edge of Crown, closer to the front

Crown Height

Distal Edge of Crown

Bite Damage or Predation Marks

Serrations

Apex

Otodus megalodon

Basic Megalodon Tooth Features. Dr. Jay M. Lipoff.

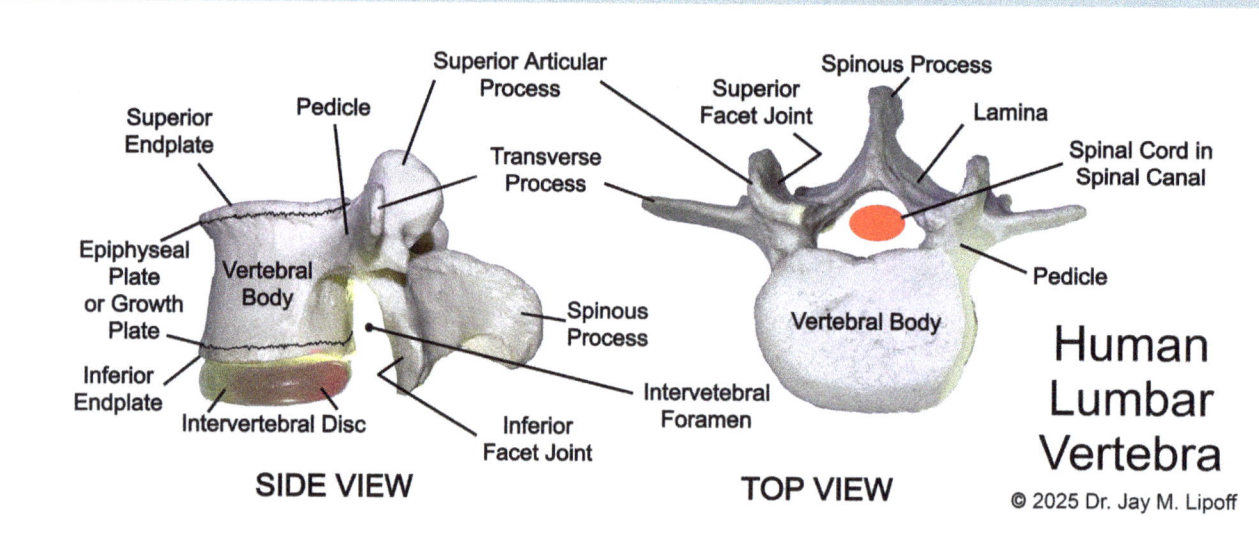

Superior Endplate

Pedicle

Superior Articular Process

Superior Facet Joint

Spinous Process

Lamina

Transverse Process

Spinal Cord in Spinal Canal

Epiphyseal Plate or Growth Plate

Vertebral Body

Pedicle

Inferior Endplate

Spinous Process

Vertebral Body

Human Lumbar Vertebra

Intervetebral Foramen

Intervertebral Disc

Inferior Facet Joint

SIDE VIEW

TOP VIEW

© 2025 Dr. Jay M. Lipoff

A Human Lumbar Vertebra to Assist in Understanding the Similarities of Parts. Dr. Jay M. Lipoff.

Megalodon Art.
Courtesy of Jaap Roos Art.

CHAPTER ELEVEN
Megalodon, Florida Fossils & More

Megalodon Images

Various images of megalodon reconstructions from around the world are followed by incredible photos related to the topics discussed in this book. There are many unbelievable contributions from dedicated fossil hunters with unreal collections and a wealth of knowledge. It has been an honor to speak with them and a privilege to include some of their findings. Scale bars were included whenever possible with all photos in this entire section.

Megalodon. (National Museum of Natural History, Smithsonian Institution, Washington, D.C.).
The model is 52 feet (~15.85 m.) in length. Photo by PLBechly on Wikimedia Commons. Licensed under CC BY-SA 4.0.

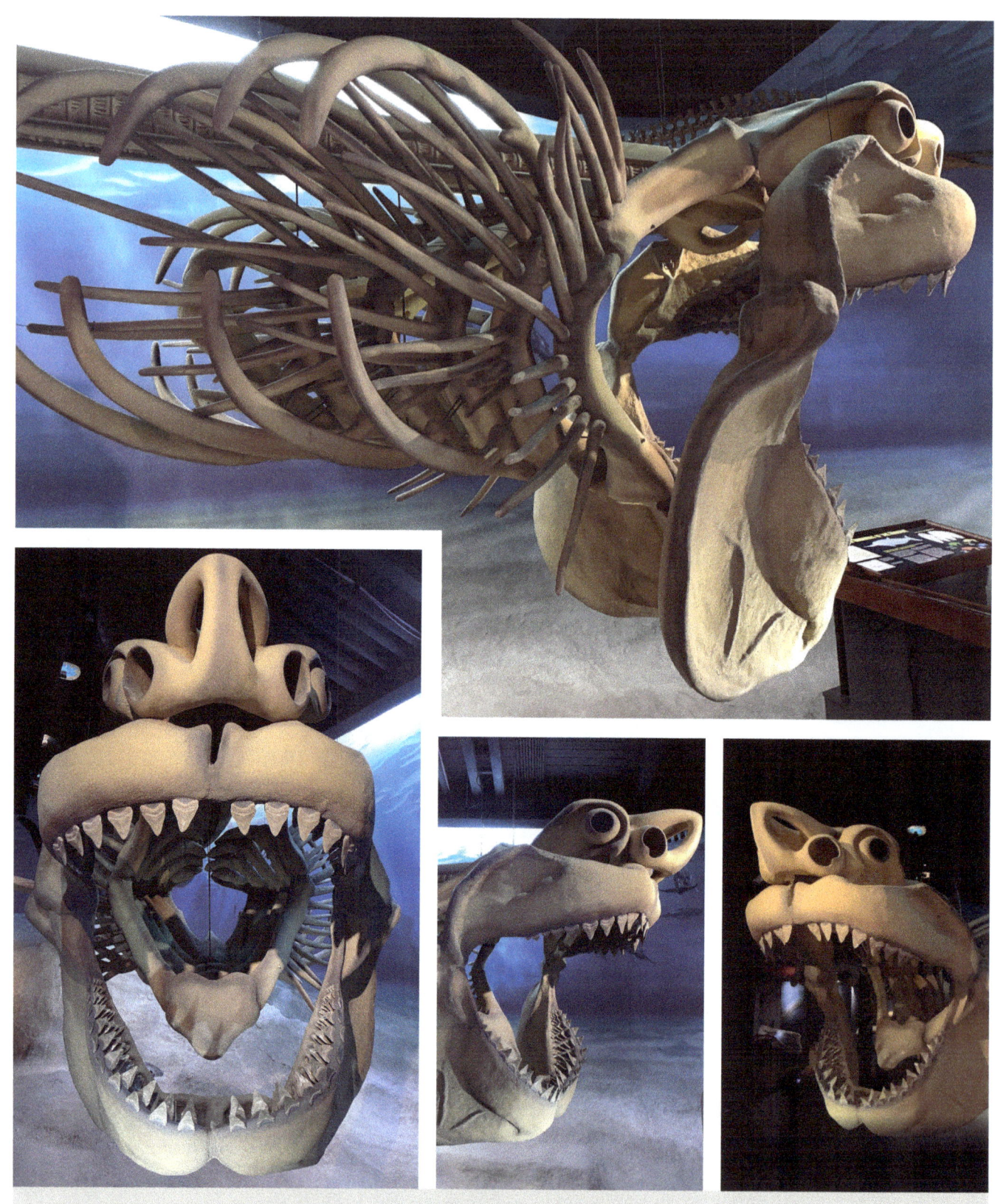

Megalodon Skeleton.
(On display at the Calvert Marine Museum, Solomons Island, MD.) Dr. Jay M. Lipoff.

Megalodon for the Natural History Museum Linz.
Photo by Werner Kraus on Wikimedia Commons. Licensed under CC BY-SA 4.0.

Life-Size Megalodon. History Museum in Puebla, Mexico.
Photo by Luis Alvaz on Wikimedia Commons. Licensed under CC BY-SA 4.0.

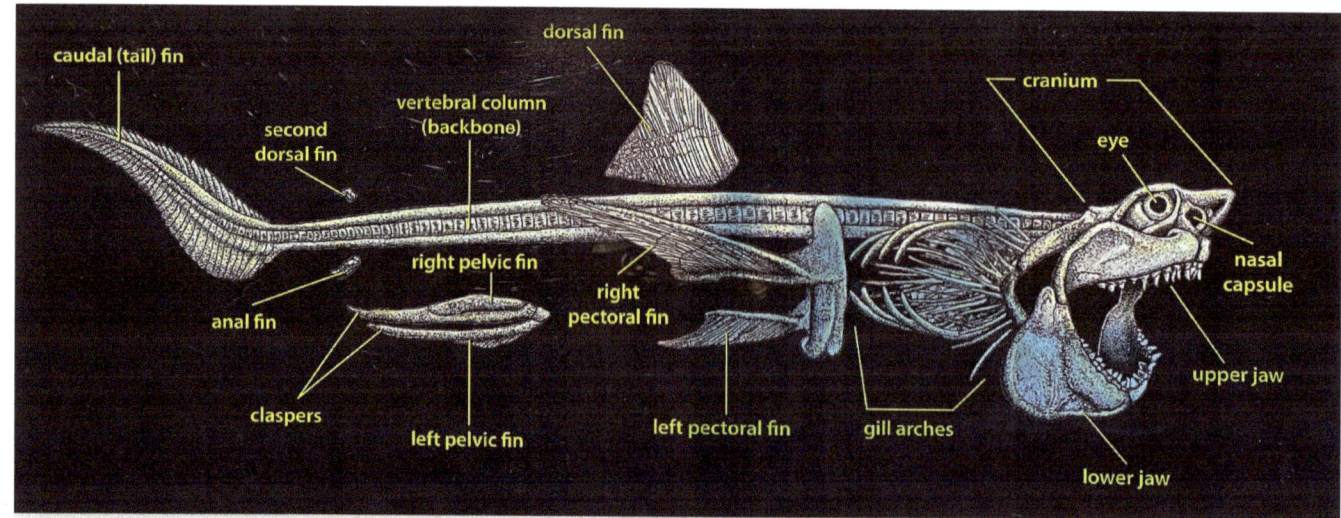

Drawing by Constance "Connie" Rankin. Courtesy of the Calvert Marine Museum.

Mote Marine Laboratory and Aquarium, Sarasota, Florida.
Courtesy of Kristina Palumbo (Nautical Necklaces, Aquanutz Scuba Diving Charters).

Megalodon Jaws

A few examples of Megalodon jaws that are on display around the world.

Either a Great White Missed its Target or Stepped on a Lego. © iStock. Credit: USO.

Bashford Dean's 1909 Reconstruction of a Megalodon Jaw for the Smithsonian Institution. © Public Domain.

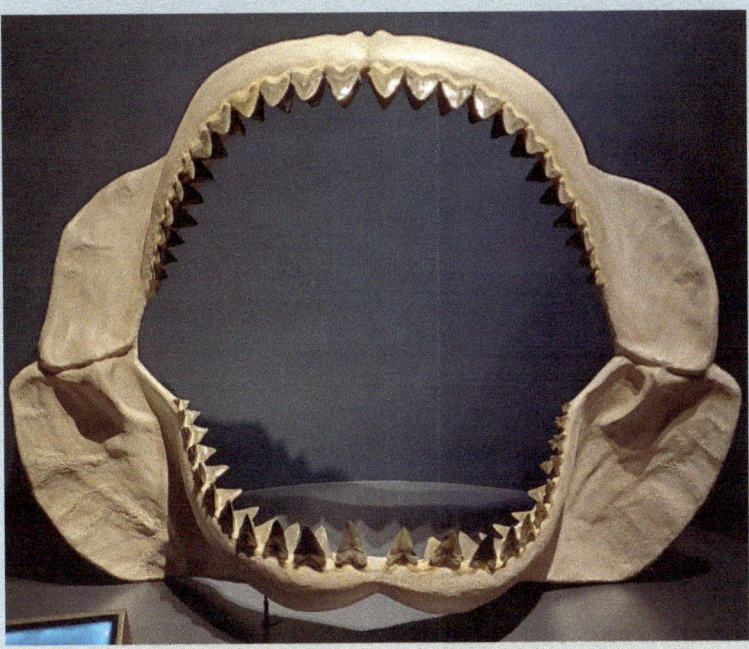

Megalodon Jaw from Tellus Science Museum. Photo by JJonahJackalop on Wikimedia Commons. Licensed under CC BY-SA 4.0.

Florida Museum of Natural History (Gainesville, FL).
Dr. Jay M. Lipoff.

**Megalodon Jaws at Buena Vista Museum
of Natural History** (Bakersfield, CA) .
Courtesy of Jason Kowinsky (fossilguy.com).

Megalodon Jaws. © Public Domain.

Megalodon at the Aurora Museum
(Aurora, NC). Dr. Jay M. Lipoff.

Megalodon Jaw. American Museum of Natural History. © Public Domain.

Shark Attack 5D Theater (Galveston, Texas). © iStock. Credit: felixmizioznikov.

Inside A Megalodon Jaw at Wyoming Fossils (Kemmerer, WY). Special thanks to Robert Bowen and Jen Edinger (Wyoming Fossils). Courtesy of Sallie Harding.

Megalodon Jaws at the Baltimore National Aquarium. Dr. Jay M. Lipoff.

Megalodon Entryway.
Courtesy of Jaap Roos Art.

**Megalodon Jaw at the Ottoneum Museum
of Natural History** (Kassel, Germany).
Courtesy of Ben and Marion Becker-Beier.

**Massive Megalodon Jaws at the Smithsonian
Museum.** Courtesy of Janet Hoard Tate.

Megalodon Teeth

Photos of megalodon teeth, because we can't get enough of them.

Courtesy of Dr. Jay M. Lipoff.

Giant Chunky Meg from Florida.
Courtesy of Aquanutz Scuba
Diving Charters.

A Pile of Megs from North Carolina.
Courtesy of Dr. Harry M. Maisch, IV.

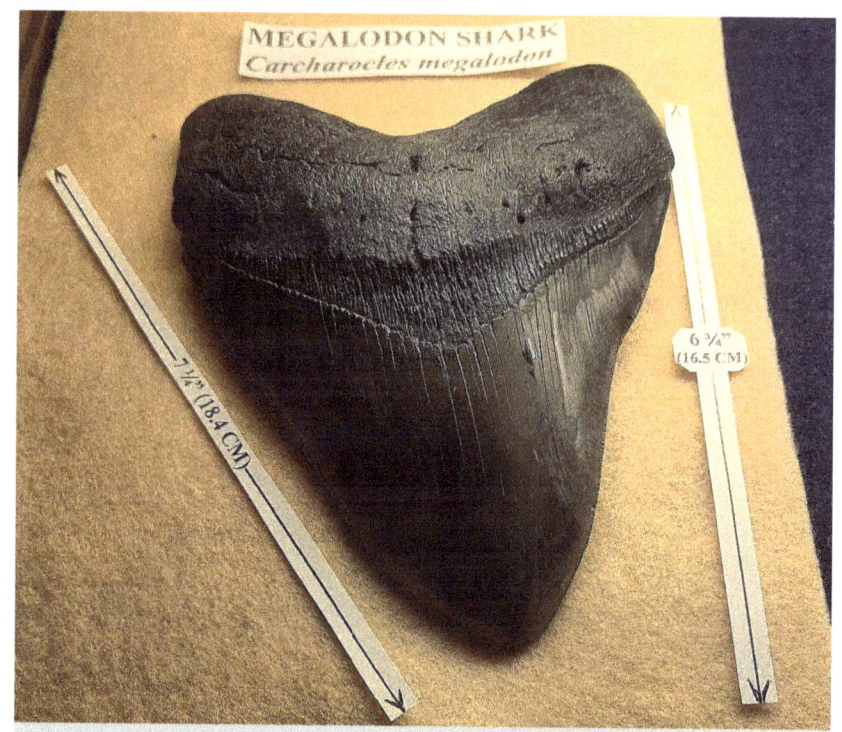

Large 7 ¼" or Bigger Megalodon Tooth
(from Dr. Gordon Hubbell's Collection).
Courtesy of Dr. Harry M. Maisch, IV.

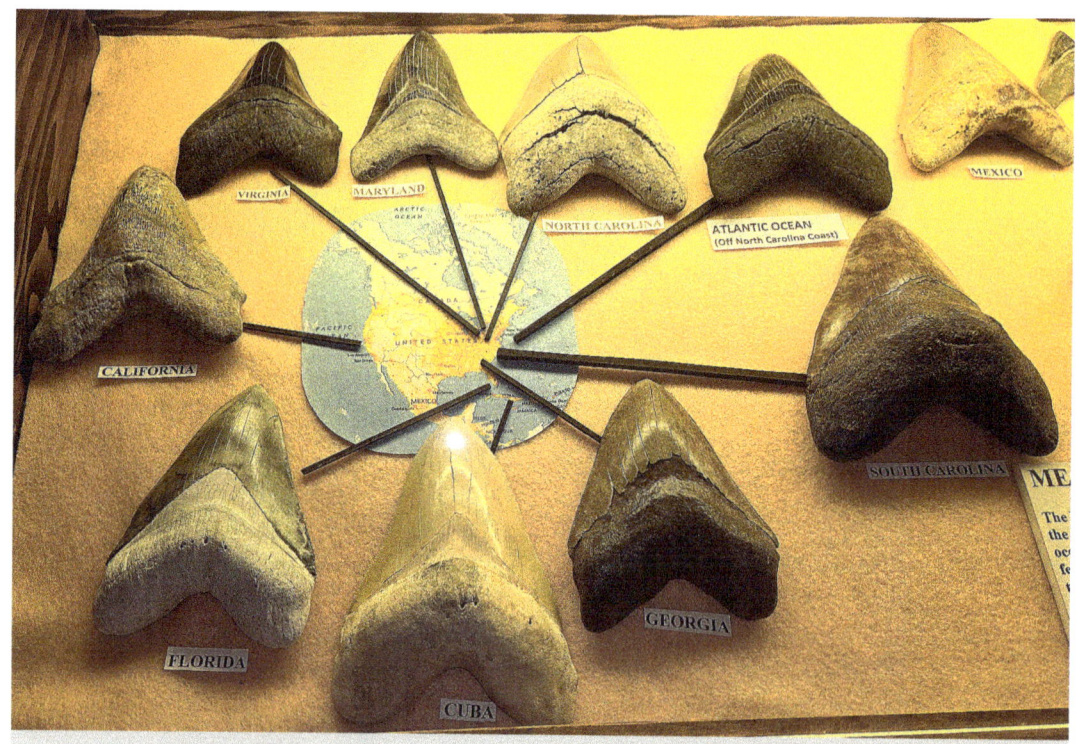

Megalodon Teeth from Around the World (from Dr. Gordon Hubbell's Collection).
Courtesy of Dr. Harry M. Maisch, IV.

Left Jaw

Upper Row of
Teeth in Left Jaw

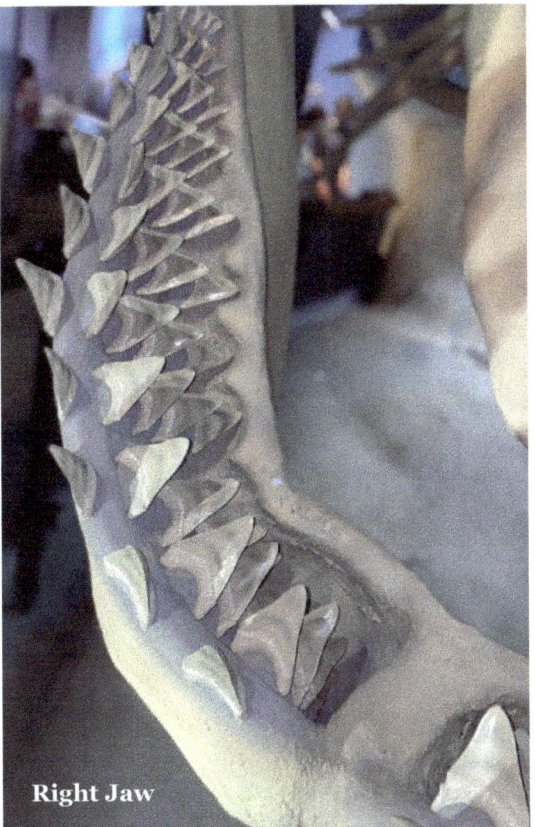

Right Jaw

Left Jaw

Teeth Inside the Mouth of *Otodus megalodon* at the Calvert Marine Museum.
Dr. Jay M. Lipoff.

Back of Tooth or Lingual Side

Front of Tooth or Labial Side

Back

Front

Two Sets of Megalodon Teeth.
Courtesy of Tara Gelsomino (Palmetto Fossil Excursions, S.C.).

Back of Tooth

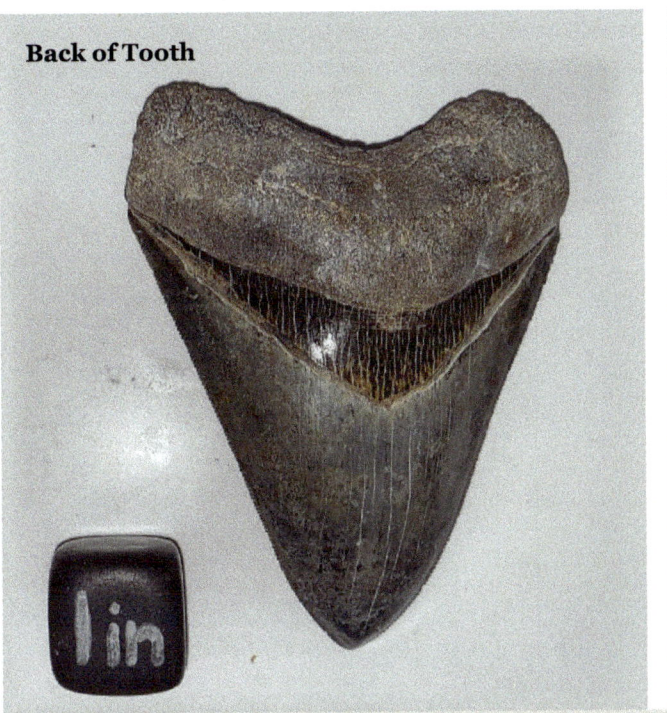

Front of Tooth

Florida Megalodon Tooth. Courtesy of Michael Konecnik (Aquanutz Scuba Diving Charters).

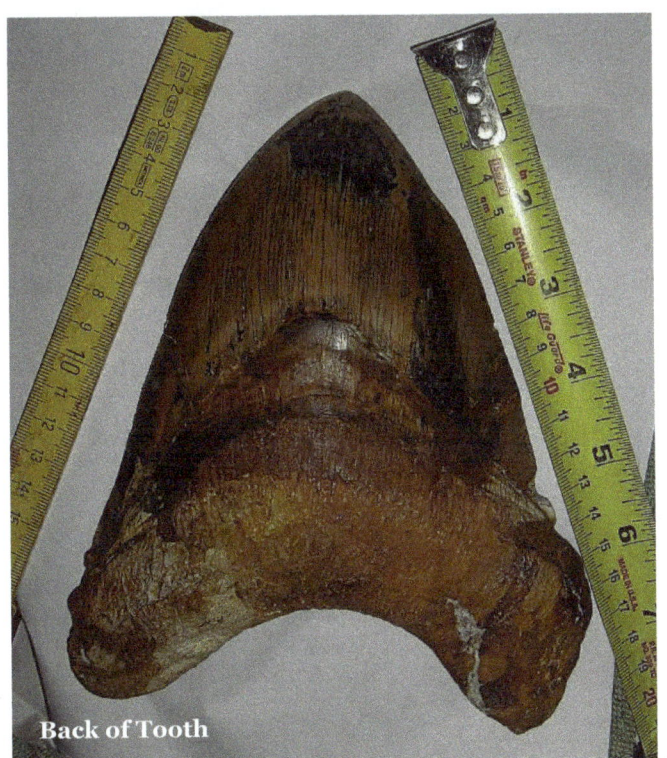

Back of Tooth

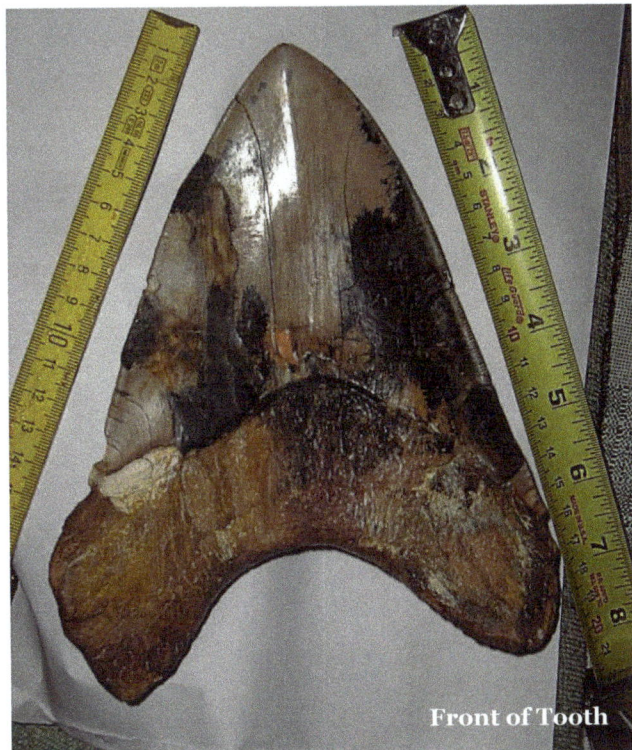

Front of Tooth

Large 7 ¼" or Bigger Megalodon Tooth from Peru. Photo by Craig Sundell.

Pathological Teeth from Megalodon's Lineage

Many examples of teeth that were not perfectly formed for one reason or another.
They are unique and rare.

Deformed Teeth (from Dr. Gordon Hubbell's Collection). Courtesy of Dr. Harry M. Maisch, IV.

Gemination of a Megalodon Tooth (left)
Compared to a Normal Megalodon Tooth.
Courtesy of Mark Kostich Photography.

**Megalodon Tooth Split into
Two Complete Teeth**
(from Dr. Gordon Hubbell's Collection).
Courtesy of Dr. Harry M. Maisch, IV.

Pathological Teeth. Courtesy of Michael Konecnik.
(Aquanutz Scuba Diving Charters).

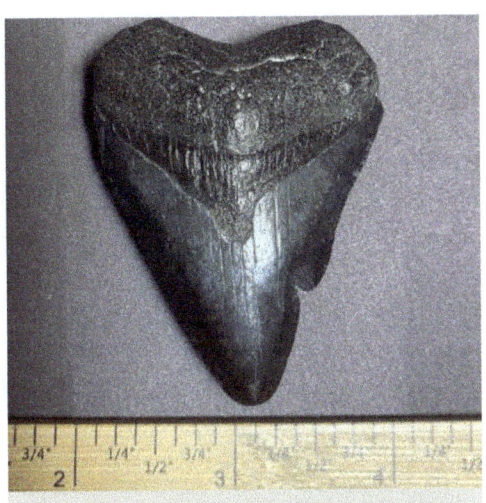

Double-Tipped Megalodon Tooth.
Courtesy of Blair Morrow.
(Meg Goddess Designs, Aquanutz).

Pathological Tooth.
Courtesy of Blair Morrow (Meg Goddess Designs, Aquanutz).

Triple-Tipped Meg Tooth.
Courtesy of Jason Soward.

Gnarly Megalodon Teeth.
Courtesy of Jason Soward (Unforgettable Oddities).

Gnarly Megalodon Teeth.
Courtesy of Michael Konecnik (Aquanutz).

Triple-Tipped Megalodon Tooth. Courtesy of Michael Nastasio.

Triple-Tipped Megalodon Tooth. Courtesy of Stephen Turner.

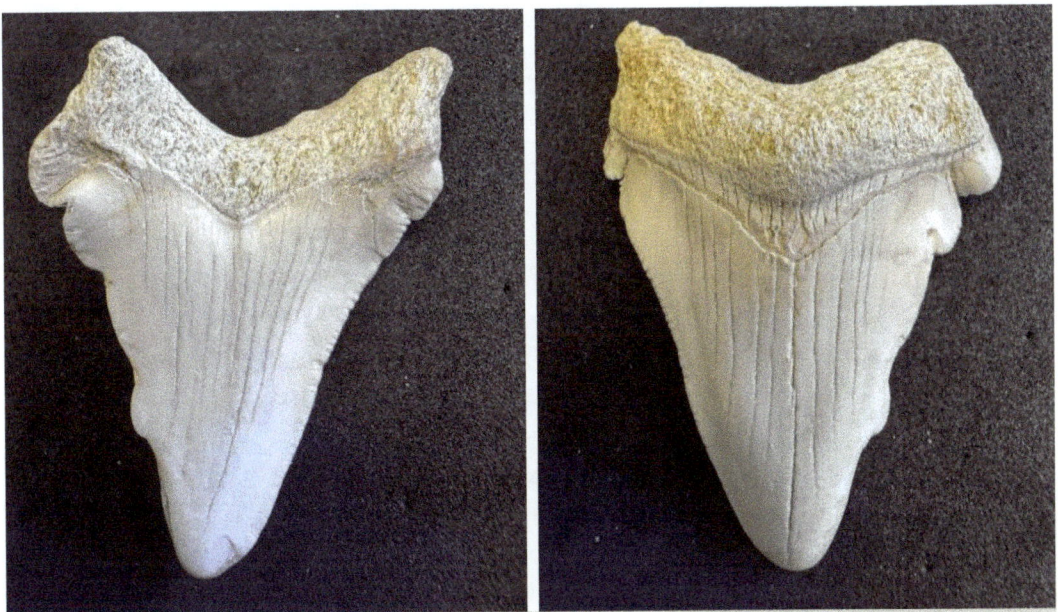

Otodus angustidens from Summerville, South Carolina. Courtesy of Larry Tanenbaum.

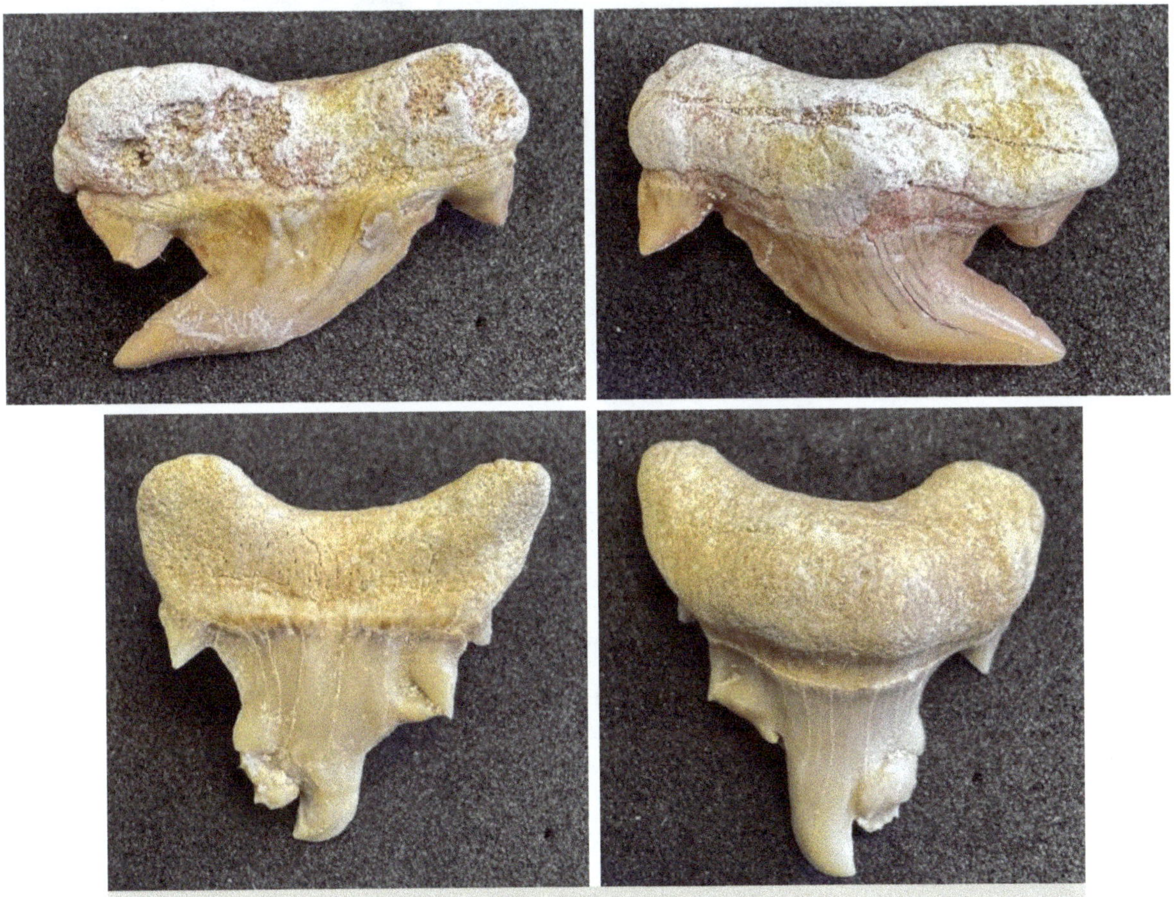

Two Sets of Pathological *Otodus obliquus* **from Morocco, Africa**.
Courtesy of Larry Tanenbaum.

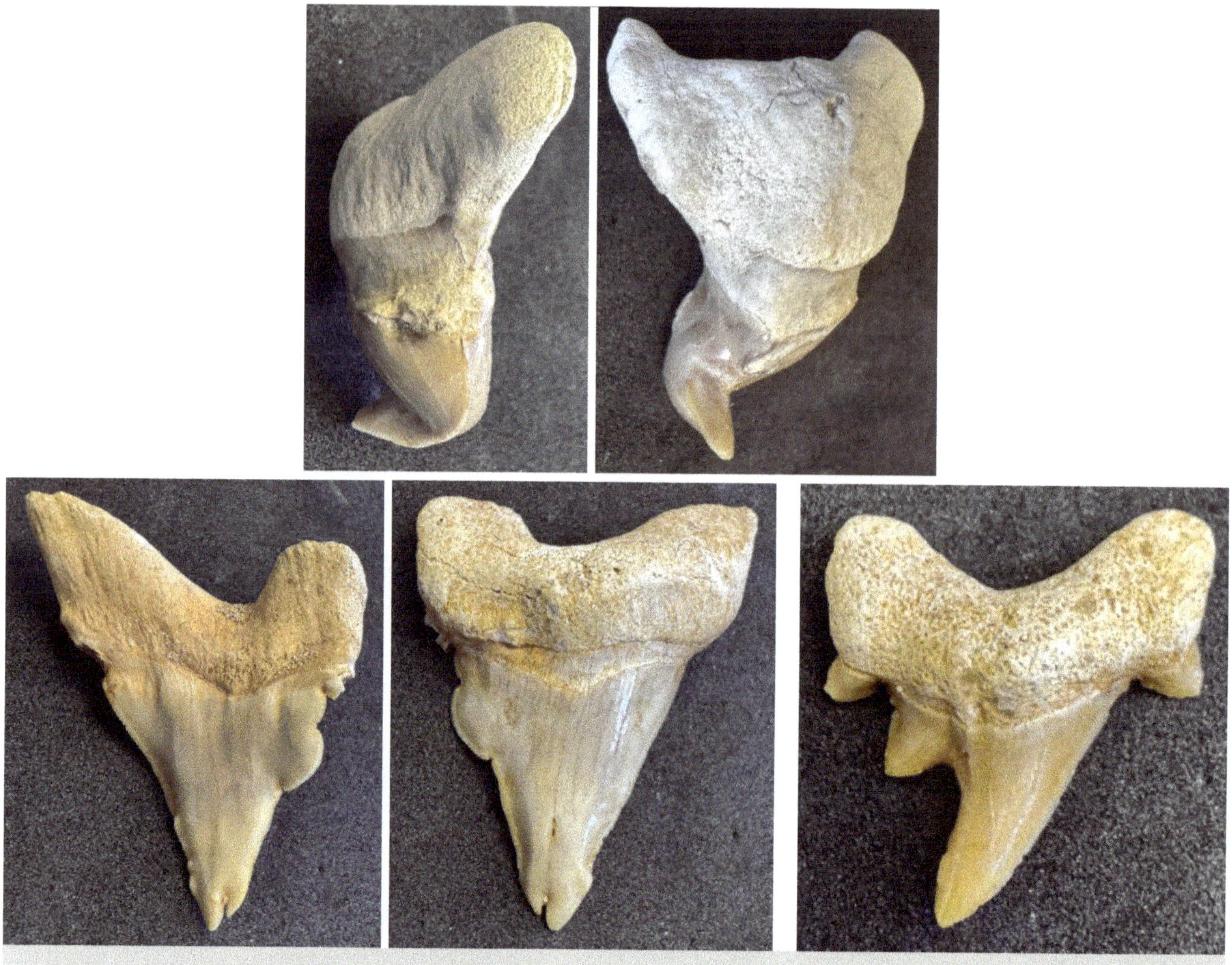

Two Sets and a Solo Pathological *Otodus obliquus* from Morocco, Africa. Courtesy of Larry Tanenbaum.

Pathological *Otodus obliquus*. Dr. Jay M. Lipoff.

Double-Tipped Meg Tooth. Courtesy of Justin Boorstein (Every Day I'm Shoveling).

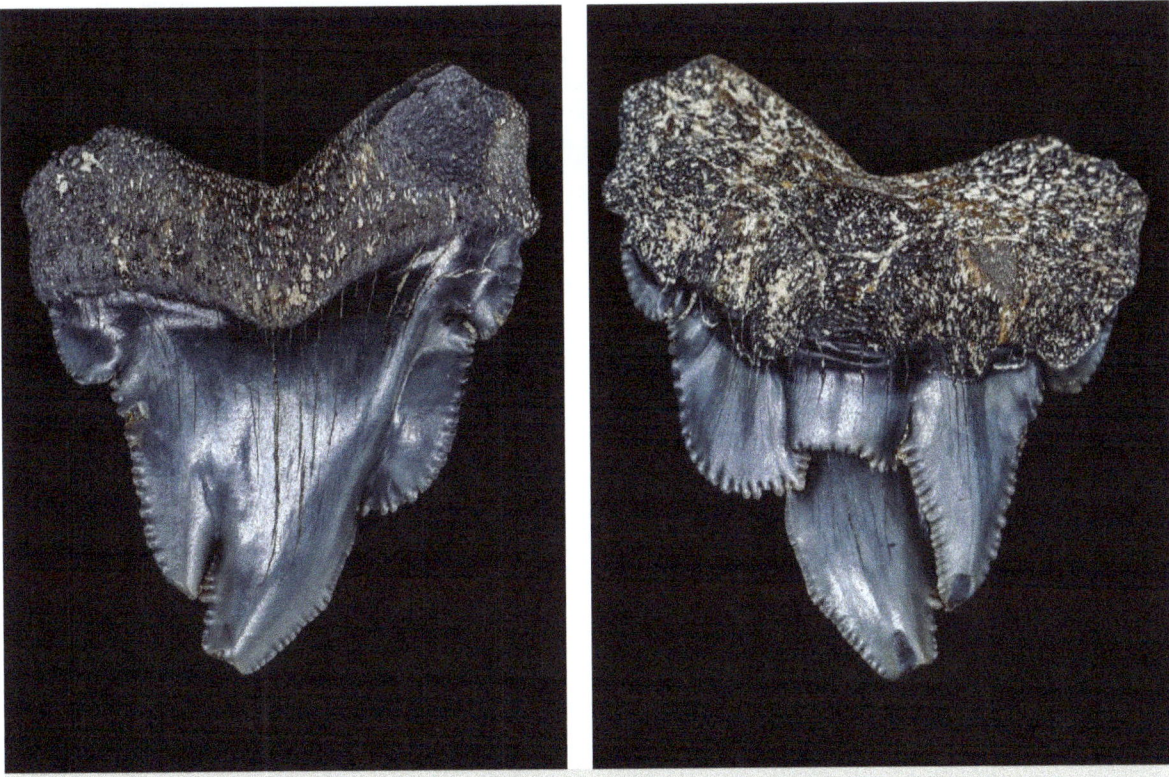

At Least a Quadruple-Tipped Megalodon Tooth. Courtesy of Dr. Adam Bozeman.

A Tooth Within a Tooth! Courtesy of Michael Konecnik (Aquanutz Scuba Diving Charters).

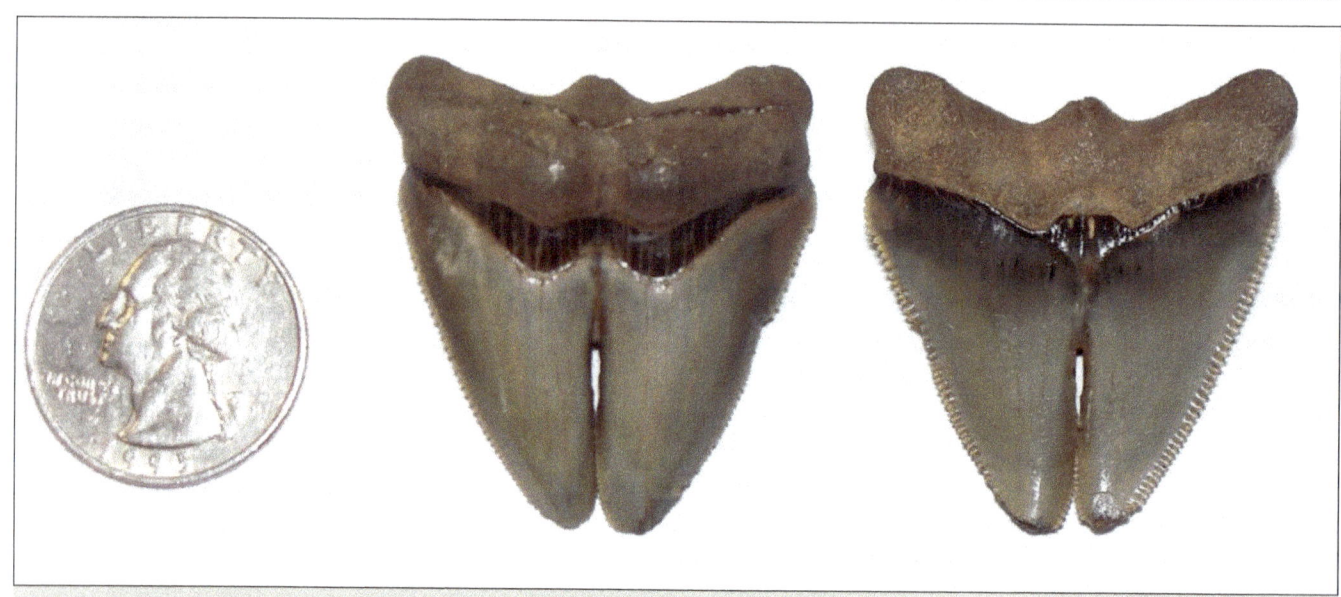

Two for One. Twin Teeth. Courtesy of Daryl Serafin.

Pathological Teeth of Other Sharks

Nobody is perfect.

Pathological Bull Shark Tooth.
Courtesy of Skylar Vertes.

Patho *Squalicorax bassanii*.
AKA a Crow Shark, from the Cretaceous Period.

Gemination in Sand Tiger Tooth. Courtesy of Mark Kostich Photography.

Lemon Shark (Gainesville, FL).
Courtesy of Larry Tanenbaum.

Lemon Shark (Gainesville, FL).
Courtesy of Larry Tanenbaum.

Lemon Shark (Orlando, FL).
Courtesy of Larry Tanenbaum.

Snaggletooth Shark (Orlando, FL).
Courtesy of Larry Tanenbaum.

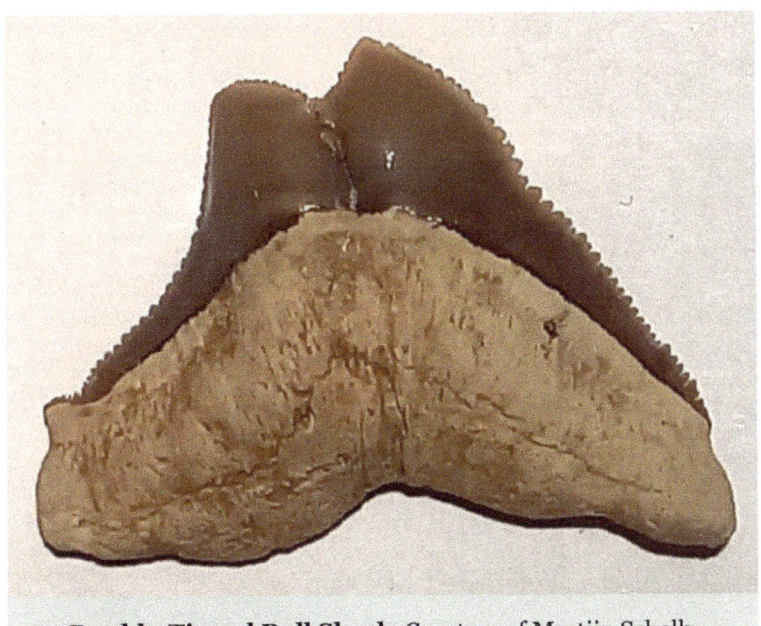

Double-Tipped Bull Shark. Courtesy of Martijn Schalk.

Double-Tipped Bull Shark Tooth.
Courtesy of Michael Konecnik.
(Aquanutz Scuba Diving Charters).

Gemination in a Bull Shark Tooth.
(Venice, FL). Courtesy of Larry Tanenbaum.

Great White Patho. Courtesy of Ryan Meyer.

Crow Shark from Morocco, Africa (Cretaceous).
Courtesy of Larry Tanenbaum.

Color Variations of Megalodon Teeth

Just when you thought Megalodon teeth couldn't get any better.

Beautiful 6.8" South Carolina Megalodon Tooth. Jason Mathias (Jason Mathias Art Studios).

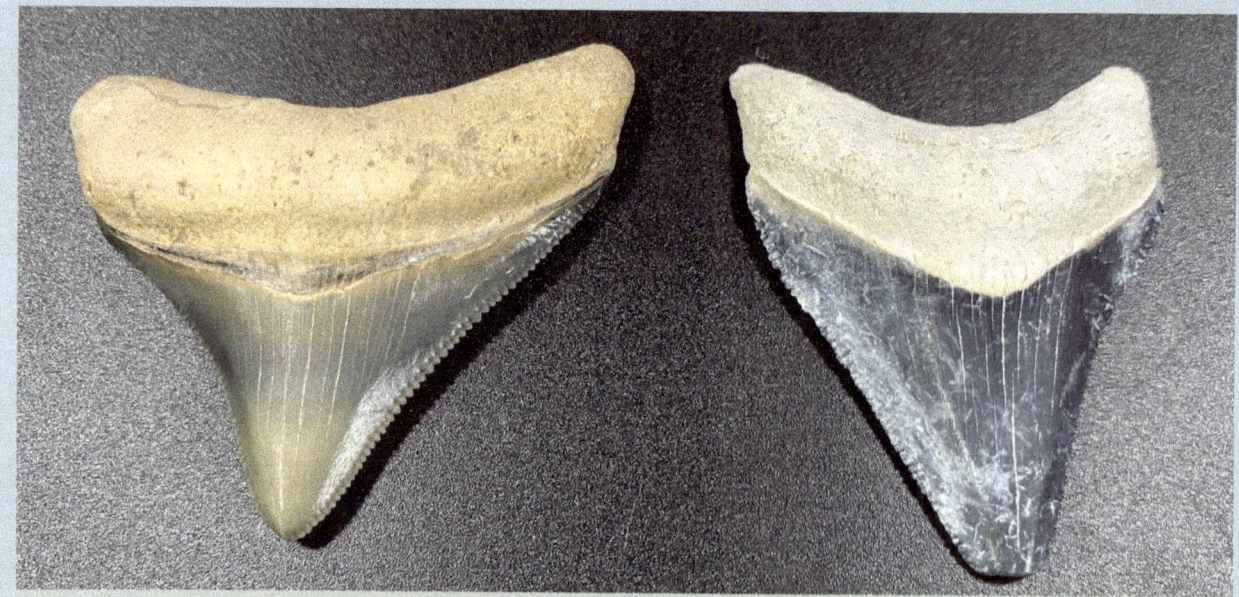

Meg Teeth with Completely Different Coloring (both are from Florida).
Golden Beach (left) and Bone Valley (right). Both are the backside views. Dr. Jay M. Lipoff.

Otodus megalodon. Courtesy of Sven Reiter.

A Large Megalodon Tooth from N.C.
Dr. Jay M. Lipoff.

Otodus megalodon. Courtesy of Adam Bozeman.

**My First Maryland Megalodon Tooth
from the Shores of the Potomac River, MD.**
Dr. Jay M. Lipoff.

Lightning Tooth of *Otodus angustidens*. Courtesy of Tara Gelsomino (Palmetto Fossil Excursion).

 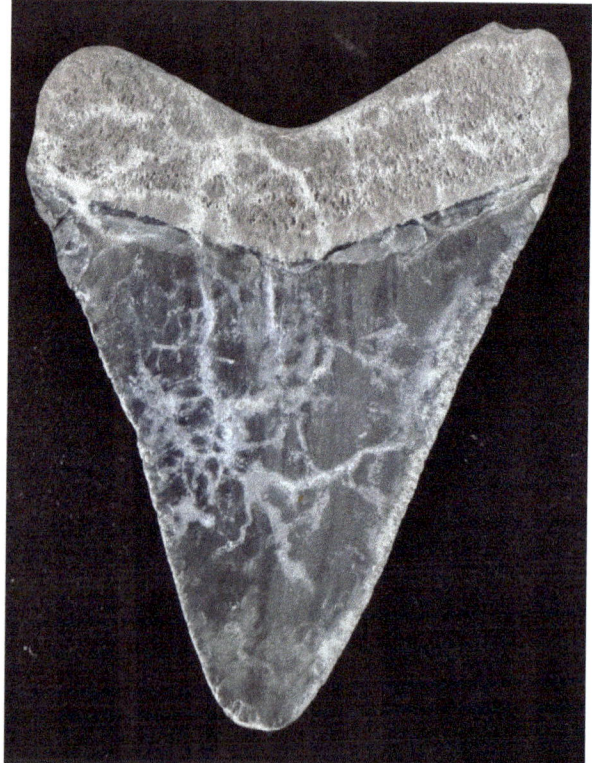

Otodus megalodon (from Bone Valley, FL). Courtesy of Larry Tanenbaum.

Otodus megalodon. Courtesy of Tommy Royal.

Otodus megalodon.
Courtesy of Dr. Adam Bozeman.

Three Beautifully Colored *Otodus megalodon* Teeth. Courtesy of Ryan Meyer.

"Hubbell Meg" or Juvenile Meg Teeth.
Courtesy of Maddy Theresa.

Serrations on a Meg.
Courtesy of Dr. Adam Bozeman.

Otodus megalodon.
Courtesy of Dr. Adam Bozeman.

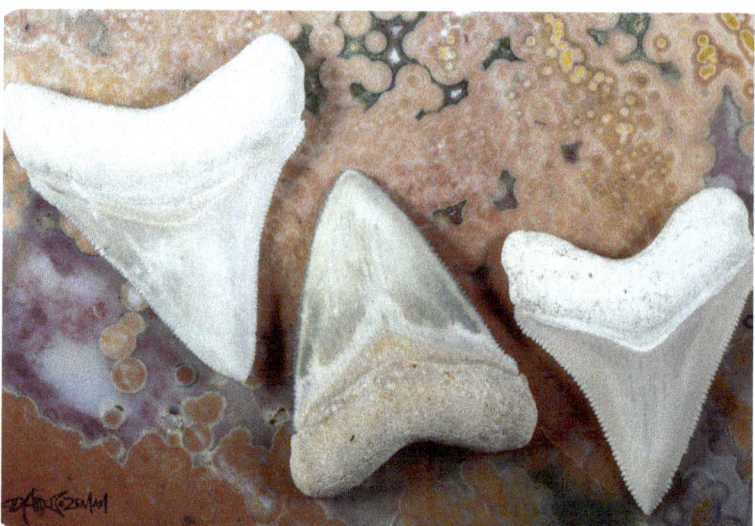

**Bone Valley White *O. megalodon*, Hubbell Meg,
and *O. chubutensis*.** Courtesy of Dr. Adam Bozeman.

Spotted Meg Tooth. Courtesy of Paul Adams.

Megs from Indonesia. Courtesy of Andrew Ensing.

Indonesian Megs. Courtesy of Dirk Rouleaux.

Indonesian Megs. Courtesy of Dirk Rouleaux.

Green Megalodon Teeth.
Courtesy of Skylar Vertes.

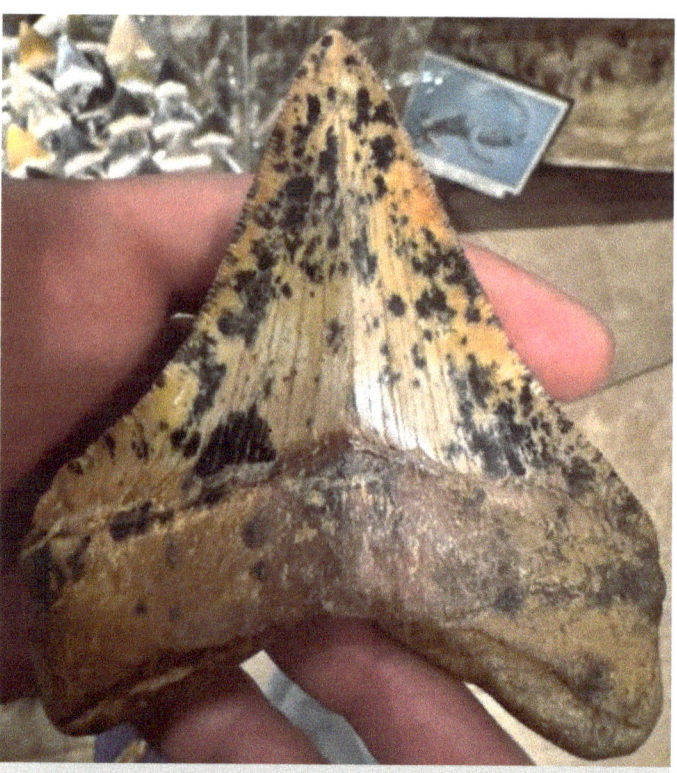

Speckled Meg Tooth. Courtesy of Sven Reiter.

Red Meg from the Meherrin River, N.C.
Courtesy of Mike Jacobsen.

"Rorschach-like," Meg from Georgia.
Courtesy of Mike Jacobsen.

Pyrite Through the Veins of this Megalodon Tooth. Courtesy of Skylar Vertes.

"Always be yourself, unless you can be Megalodon. Then be Megalodon. That is, of course, unless you can be BatMeg." Courtesy of Dr. Adam Bozeman.

Color Variations in Other Species

So much beautiful color in these happy little teeth that
even Bob Ross would have been impressed.

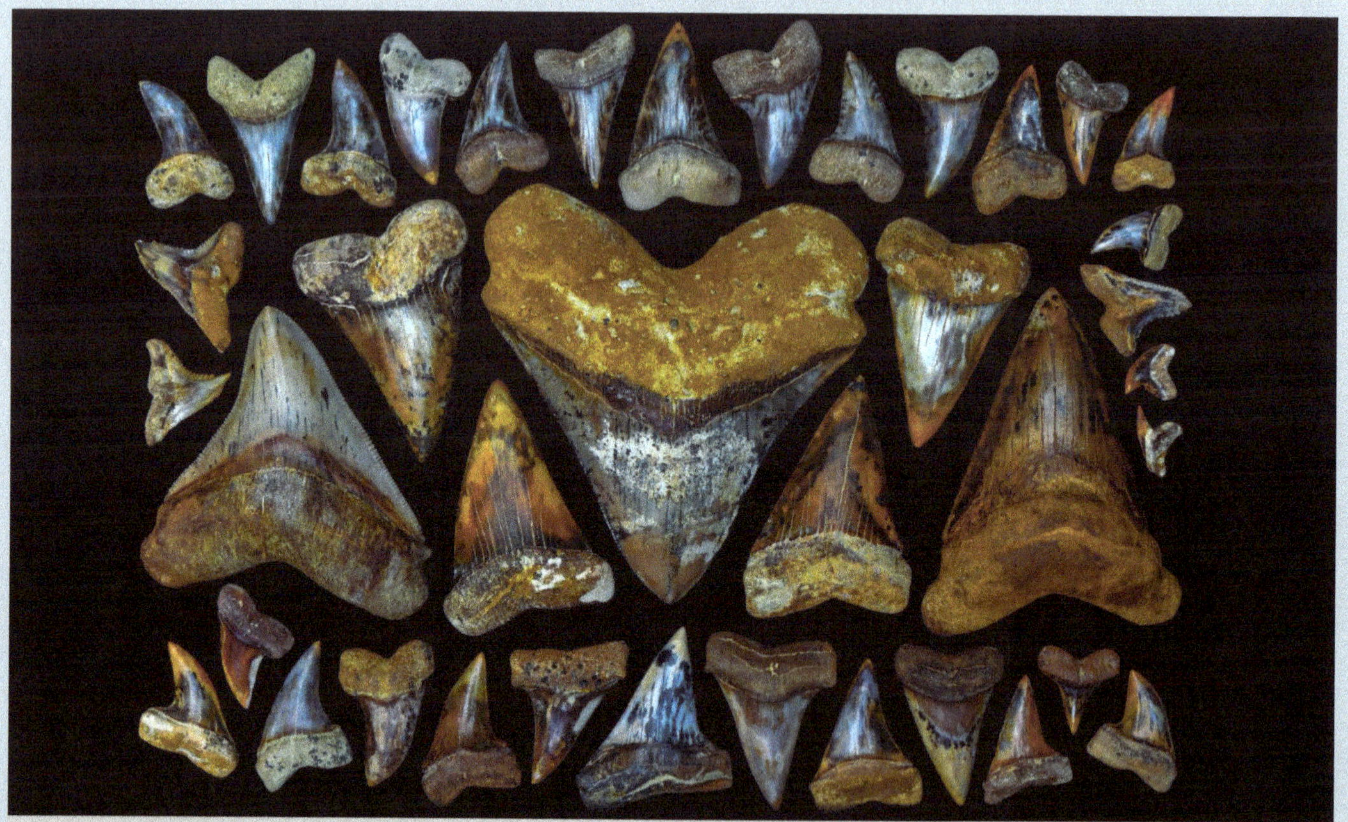

Courtesy of Dr. Adam Bozeman.

Lightning Mastodon Tooth.
Courtesy of Joseph Branin.

Carcharodon hastalis or **Extinct Lesser
White Shark Teeth.** Courtesy of Dr. Adam Bozeman.

***Hemipristis serra* or Snaggletooth Shark Tooth.**
Courtesy of Bonnie Nowell.

***Carcharodon carcharias* or Great White.**
Courtesy of Tara Gelsomino.
(Palmetto Fossil Excursions).

***Carcharodon planus* (front).**
Courtesy of Jeff Lefebvre.

***Carcharodon planus* (back).**
Courtesy of Jeff Lefebvre.

Galeocerdo cuvier or Tiger Shark Teeth. Courtesy of Dan Case.

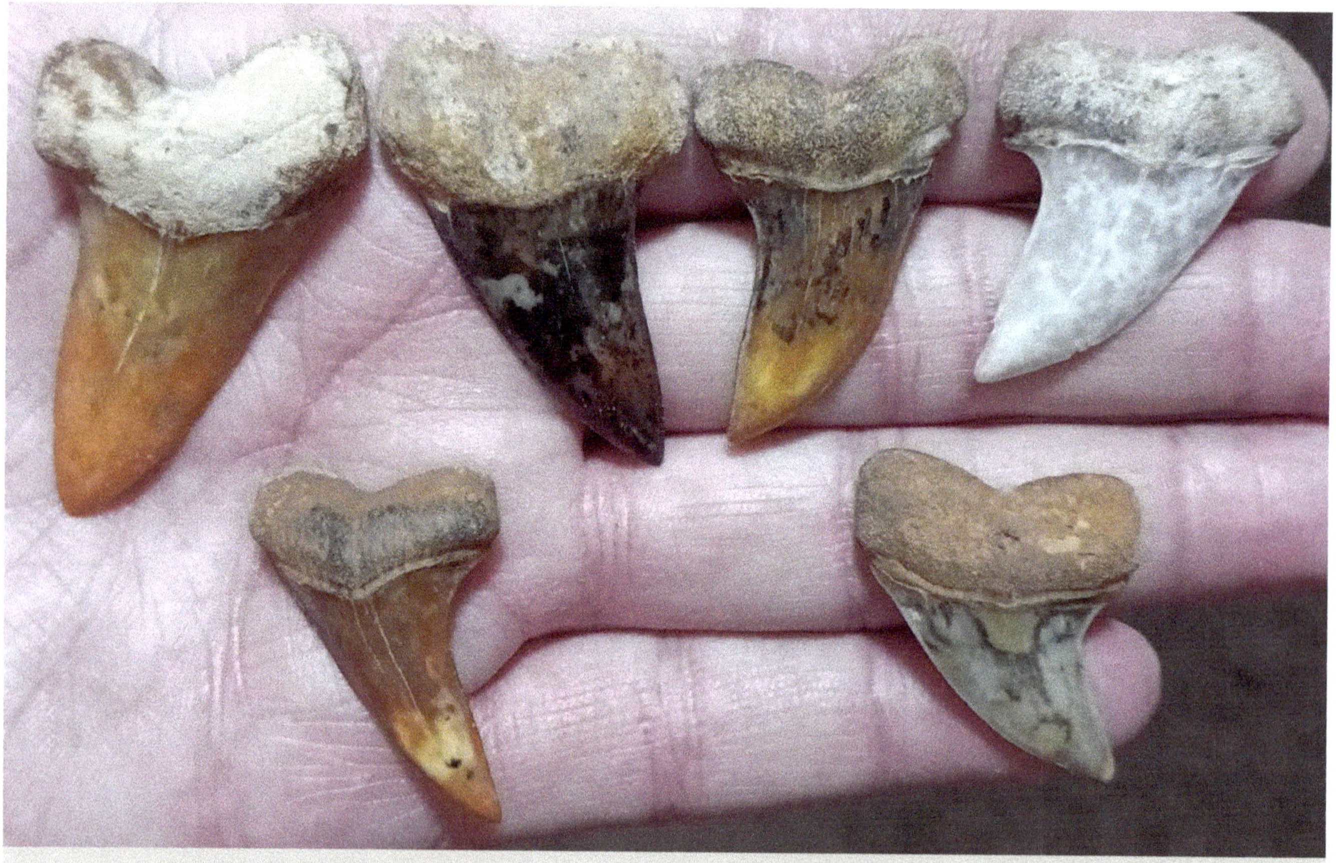

Carcharodon planus. Came after hastalis in white shark evolution, but went extinct. Courtesy of Dan Case.

Carcharodon hastalis **or Extinct Lesser White Shark Teeth**. Courtesy of Dan Case.

Hemipristis serra **or Snaggletooth Shark Teeth**. Courtesy of Dan Case.

***Carcharodon carcharias* or Great White Shark Teeth.**
Courtesy of Dan Case.

***Carcharodon carcharias* or Great White Shark Teeth.**
(Jacksonville, FL) Courtesy of Larry Tanenbaum.

***Carcharodon hubbelli* (Transitional White Shark) from Chile**. Courtesy of Larry Tanenbaum.

***Carcharodon hastalis* (Lesser White), Bakersfield, California.**
(Left two) Courtesy of Larry Tanenbaum. (Right) Courtesy of Sven Reiter.

Carcharodon hastalis **or Extinct Lesser White Shark Teeth**.
Courtesy of Stephen Lee Wenzel (SWFL Fossil Discovery's).

Great White.
Courtesy of Stephen Lee Wenzel.
(SWFL Fossil Discovery's).

Great White.
Courtesy of Ryan Meyer.

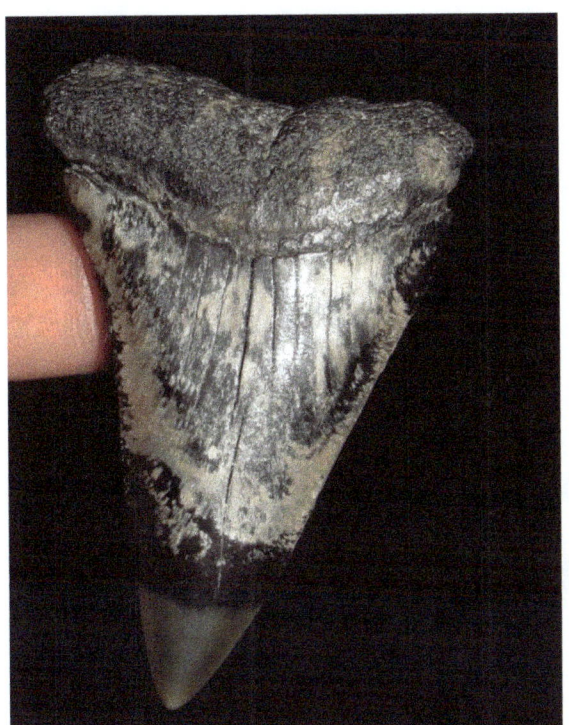

Carcharodon hastalis **or Extinct Lesser White Shark Teeth.** Courtesy of Ryan Meyer.

Hemipristis serra **or Snaggletooth Shark Teeth.**
Courtesy of Dr. Adam Bozeman.

Galeocerdo cuvier* or *Tiger Shark.
Courtesy of Dr. Adam Bozeman.

Great White. Courtesy of Dr. Adam Bozeman.

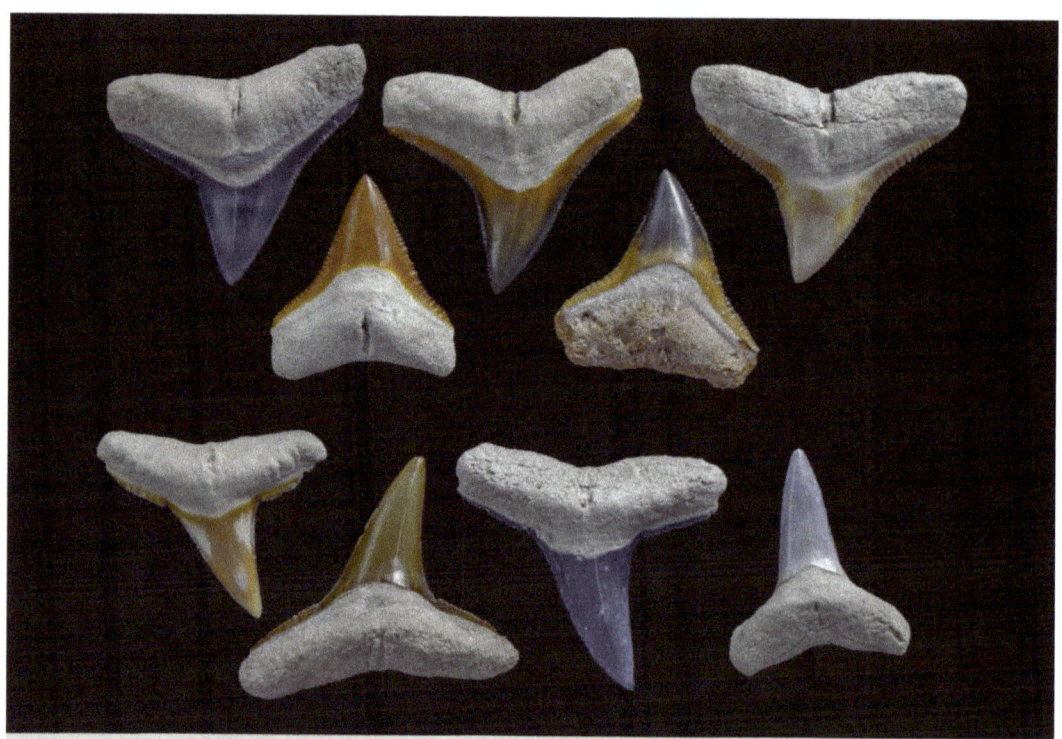

Bull Shark (5, top row) **vs. Lemon Shark** (4, bottom row).
Bone Valley, Florida. Courtesy of Mark Kostich (Mark Kostich Photography).

Hemipristis serra or **Snaggletooth Shark**. Courtesy of Skylar Vertes.

Carcharodon hastalis or **Extinct Lesser White Shark Teeth**. Courtesy of Jeff Lefebvre.

Australian Great White Teeth. Courtesy of Larry Tanenbaum.

A White Great White Tooth.
Courtesy of Mike Jacobson.

A Chocolate Great White Tooth.
Courtesy of Skylar Vertes.

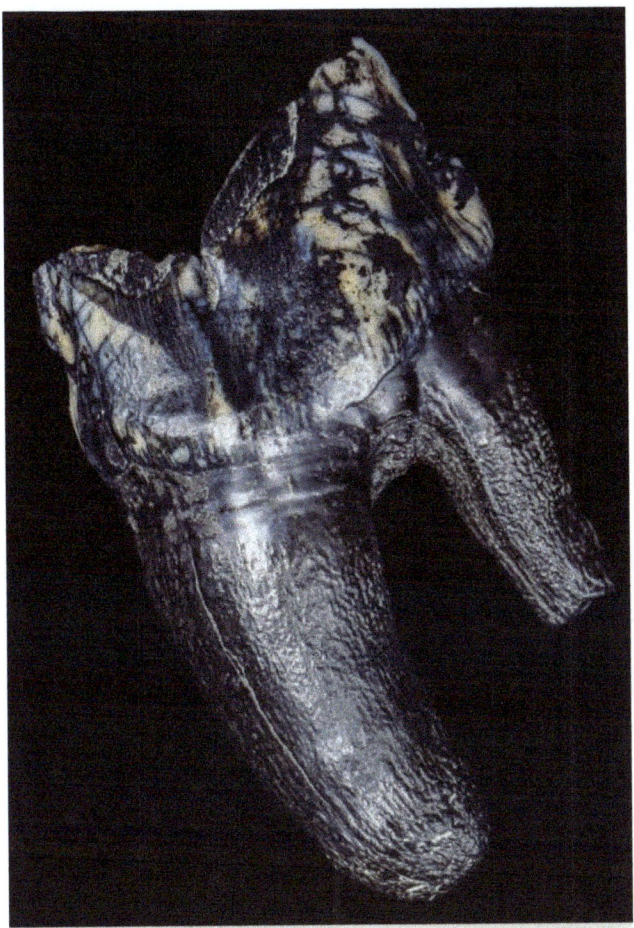

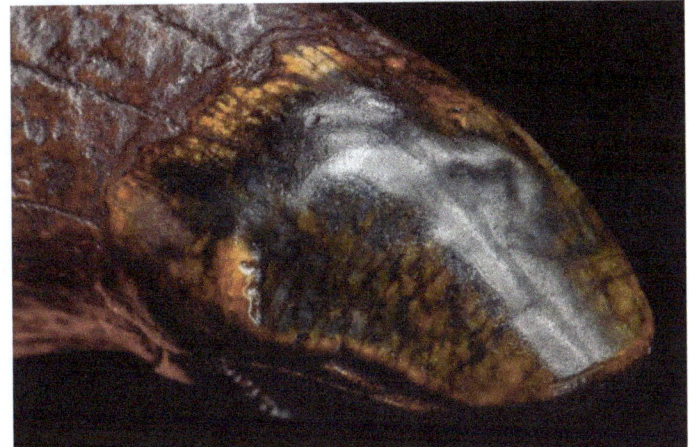

Smilodon fatalis. Courtesy of Josh Galloway.
Photos taken by Dr. Adam Bozeman.

***Panthera atrox,* or The American Lion**.
Courtesy of Josh Galloway.
Photos taken by Dr. Adam Bozeman.

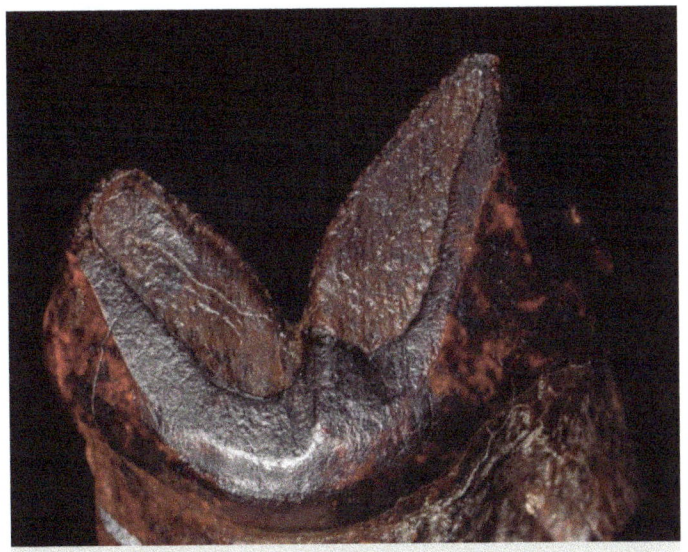

***Hexanchus griseus* Shark, or Bluntnose Sixgill
Shark, Belgium.** Courtesy of Josh Galloway.
Photos taken by Dr. Adam Bozeman.

***Panthera onca,* or Jaguar**. Courtesy of Josh Galloway.
Photos taken by Dr. Adam Bozeman.

Hexanchus andersoni, or Cow Shark. Shark Tooth Hill, near Bakersfield, CA. Courtesy of Josh Galloway. Photos taken by Dr. Adam Bozeman.

Notorynchus primigenius. Sevengill Shark. Courtesy of Josh Galloway. Photos taken by Dr. Adam Bozeman.

Hexanchus griseus, Peru. Courtesy of Josh Galloway. Photos taken by Dr. Adam Bozeman.

Hexanchus griseus, Symphyseal Tooth, Peru. Courtesy of Josh Galloway. Photos taken by Dr. Adam Bozeman.

Hexanchus griseus, Chile. Courtesy of Josh Galloway. Photos taken by Dr. Adam Bozeman.

Notorynchus primigenius, Germany. Courtesy of Josh Galloway. Photos taken by Dr. Adam Bozeman.

Notorynchus kempi, Kazakhstan. Courtesy of Josh Galloway. Photos taken by Dr. Adam Bozeman.

CHAPTER TWELVE

Florida Fossils & Much More

This is a picture of what you might find off the coast of Venice, Florida, in the rivers, and on land. I made several dives and brought everything back to share with kids at school talks. Most of the examples in this section will be from Florida; however, I will occasionally use a great picture from another region because it provides a clearer or more complete representation of the fossil.

Dive Vacation in Venice. Dr. Jay M. Lipoff.

What Do We Have Here?

All photos taken by Blair Morrow and courtesy of Blair and Michael Konecnik
(Aquanutz Scuba Diving Charters).

Courtesy of Blair Morrow and Michael Konecnik.
(Aquanutz Scuba Diving Charters).

Courtesy of Blair Morrow and Michael Konecnik.
(Aquanutz Scuba Diving Charters).

Courtesy of Blair Morrow and Michael Konecnik.
(Aquanutz Scuba Diving Charters).

Captain Mike Konecnik. He's a fossil magnet underwater.
Pictures courtesy of Blair Morrow (Meg Goddess Designs,
Aquanutz Scuba Diving Charters).

Shark Teeth

BLACK GOLD

Meg from the Gulf near Venice, FL.
Courtesy of Mike Nastasio.

AQUANUTZ

Meg from the Gulf near Venice, FL.
Courtesy of Michael Konecnik.
(Aquanutz Scuba Diving Charters).

Parotodus benedini,
A Very Rare False Mako Shark.
Courtesy of Michael Konecnik.
(Aquanutz Scuba Diving Charters).

AQUANUTZ

Two Meg Teeth One in Matrix or Encrusted (left).
Courtesy of Michael Konecnik (Aquanutz Scuba Diving Charters).

Beautiful Megalodon Tooth with Serrations.
Courtesy of Michael Konecnik.
(Aquanutz Scuba Diving Charters).

Three Meg Teeth of Various Levels of Preservation and Completeness.
These are from shore and boat dives. Dr. Jay M. Lipoff.

Carcharodon carcharias. **Upper and Lower**
(back or lingual side). Courtesy of Michael Konecnik.
(Aquanutz Scuba Diving Charters).

Carcharodon carcharias. **Upper and Lower**
(front or labial side). Courtesy of Michael Konecnik.
(Aquanutz Scuba Diving Charters).

Thresher Shark. Courtesy of Dean Rogers.

Columbian Mammoth Fossils

Columbian Mammoth Tooth.
Courtesy of Michael Konecnik.
(Aquanutz Scuba Diving Charters).

Mammoth Jaw with Teeth.
Courtesy of Michael Konecnik.
(Aquanutz Scuba Diving Charters).

Mammoth Tooth Complete.
Courtesy of Michael Konecnik (Aquanutz Scuba Diving Charters).

Juvenile Mammoth. Courtesy of Justin Boorstein.
(Every Day I'm Shoveling).

Mammoth Tooth Found.
Courtesy of Michael Konecnik.
(Aquanutz Scuba Diving Charters).

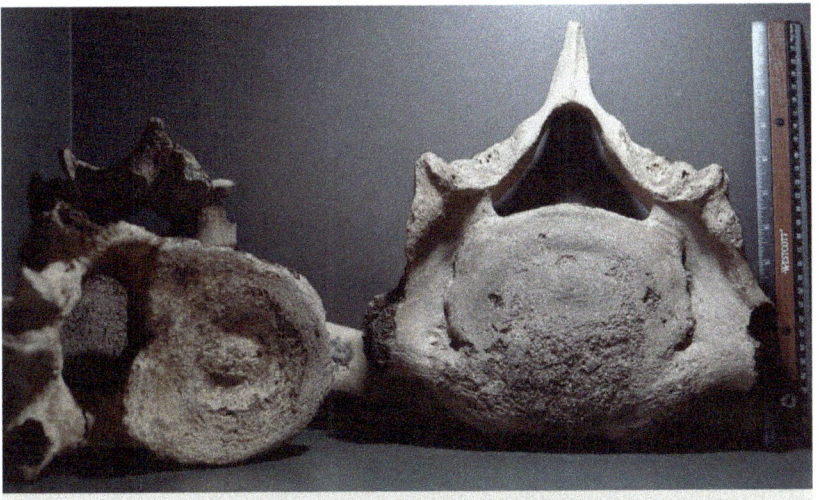

Columbian Mammoth Vertebrae.
Courtesy of Michael Tyler Staab (I Hunt Dead Things).

Captain Michael Konecnik Never Returns Empty-Handed. He has an Eagle Eye.
Courtesy of Michael Konecnik (Aquanutz Scuba Diving Charters) and Blair Morrow (Meg Goddess Designs, Aquanutz).

Cleaning Up the Treasures. Courtesy of Michael Konecnik (Aquanutz Scuba Diving Charters) and Blair Morrow (Meg Goddess Designs, Aquanutz).

It's Called a Spit Tooth. Discarded as a new tooth replaces it. Courtesy of Michael Konecnik (Aquanutz Scuba Diving Charters) and Blair Morrow (Meg Goddess Designs, Aquanutz).

Columbian Mammoth Tusk.
Courtesy of Alex Lundberg. He just found another one, too.

Columbian Mammoth Tusk.
Courtesy of Dr. Harry M. Maisch, IV.

Blair Morrow's Columbian Mammoth Tusk.
They used zip ties to keep it together while recovering it. She just located another one as well.
Courtesy of Blair Morrow (Meg Goddess Designs and Aquanutz Scuba Diving Charters).

Horse Teeth

Horse Teeth in Jaw. Courtesy of Michael Tyler Staab (I Hunt Dead Things).

Horse Incisor.
Courtesy of Michael Konecnik.
(Aquanutz Scuba Diving Charters).

Horse Teeth in Jaw and Solo Tooth.
Courtesy of Michael Konecnik.
(Aquanutz Scuba Diving Charters).

Canon Bone, Astralgus, Incisor & Molar.
Courtesy of Blair Morrow (Meg Goddess Designs
and Aquanutz Scuba Diving Charters).

A Great Day of Diving.
Courtesy of Blair Morrow (Meg Goddess Designs and
Aquanutz Scuba Diving Charters).

Phalanx.
Courtesy of Blair Morrow (Meg Goddess Designs
and Aquanutz Scuba Diving Charters).

Horse Teeth. Dr. Jay M. Lipoff.

Mastodon Fossils

Mastodon Tooth with Root, from OK. Courtesy of Garret Hernandez.

Mastodon Lightning Tooth with Roots, from FL. Courtesy of Joseph Branin.

Baby Mastodon Teeth, found together in North Florida. Courtesy of Josh Galloway.

Mastodon Teeth and Jaw, from a Florida River.
Pictures are courtesy of Joseph Branin.

Mastodon Tusk with Schreger Lines in Cross Section.
Courtesy of Michael Konecnik (Aquanutz Scuba Diving Charters).

Mastodon Tooth.
Courtesy of Michael Konecnik
(Aquanutz Scuba Diving Charters).

Mastodon Tusk Sections and Teeth.
Courtesy of Michael Tyler Staab (I Hunt Dead Things).

Predation Marks

Meg-Bitten Cetacean Vertebra.
Courtesy of Dr. Stephen J. Godfrey.

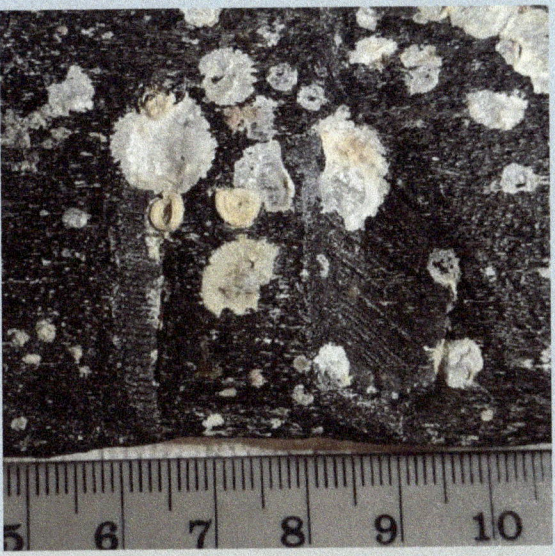

Whale Rib with Megalodon Bite Marks.
Courtesy of Dr. Harry M. Maisch, IV.

Teeth Grooves atop Turtle Shell.
Courtesy of Aquanutz Scuba Diving Charters.

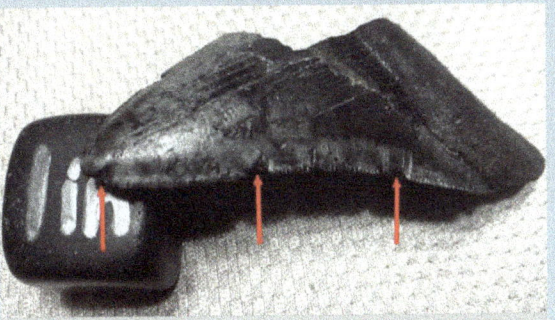

**Predation Damage to the
Edge of the Meg**. Dr Jay M. Lipoff.

Predation Marks. Courtesy of Michael Konecnik (Aquanutz Scuba Diving Charters).

Self-Bitten Megalodon Tooth.
Courtesy of Mitchell Glenn Winter.

Bite Marks on a Whale Vertebra.
Calvert Marine Museum Display. Dr. Jay M. Lipoff.

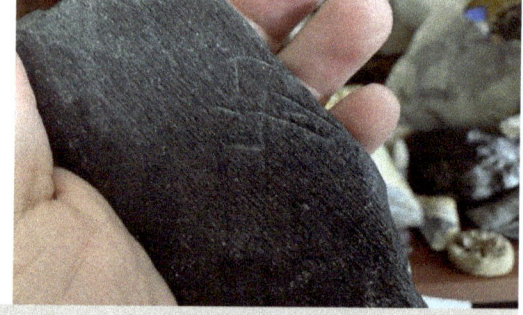

Turtle Shell and Whale Rib with Some Minor Bite Damage. Dr. Jay M. Lipoff.

Bite Marks on a Meg Tooth. Courtesy of Blair Morrow.
(Meg Goddess Designs and Aquanutz Scuba Diving Charters).

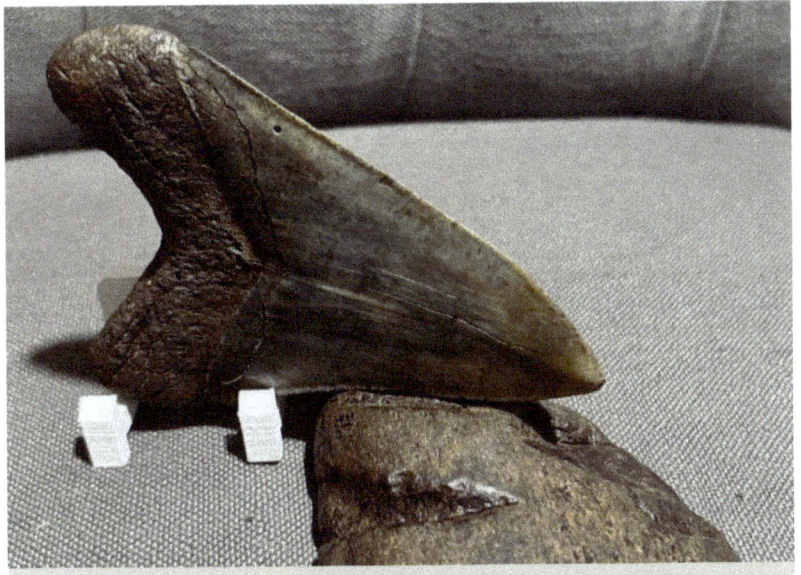

Deep Gouges in Whale Rib. The Megalodon tooth was placed there for reference. Courtesy of Nathan Foster.

Megalodon Tooth Embedded in Vertebra. Placed There by Nature.
Courtesy of Blair Morrow (Meg Goddess Designs and Aquanutz Scuba Diving Charters).

Vertebrae

Various Stages of Shark Vertebrae Erosion. Dr. Jay M. Lipoff.

MEGATOOTH SHARK VERTEBRAE
Otodus megalodon

Meg Vertebrae (on display at Calvert Marine Museum, Maryland). Dr. Jay M. Lipoff.

Vertebrae from Florida, Virginia, and Maryland.
Dr. Jay M. Lipoff.

Atlas Vertebra on Axis.
They are not associated (above & below).
Dr. Jay M. Lipoff.

(left) **Separated Epiphyseal Plate and Juvenile Vertebra with Two Visible Epiphysis Lines or Growth Plates** (next to it). **The Top Plate is Separated from the Vertebra** (right). These are found and described as "cookies." They can be 6" (~15.24 cm) across, or bigger. Calvert Cliffs, Maryland. Dr. Jay M. Lipoff.

Other Cool Finds

Alligator Scute or Osteoderm. Courtesy of Michael Tyler Staab (I Hunt Dead Things).

Alligator Vertebrae. Courtesy of Michael Tyler Staab (I Hunt Dead Things).

Alligator Teeth in Jawbone. Courtesy of Michael Tyler Staab (I Hunt Dead Things).
Alligator Tooth (inset upper right). Courtesy of Michael Konecnik (Aquanutz Scuba Diving Charters).

Bison Jaw with Teeth. Courtesy of Alex Lundberg.

Horse Jaw and Teeth. Courtesy of Dean Rogers.

Capybara Tooth. Courtesy of Michael Konecnik (Aquanutz Scuba Diving Charters).

Coprolite. Courtesy of Justin Boorstein (Every Day I'm Shoveling).

Cow Shark Tooth.
Shark Tooth or Hollis Island, VA. Dr. Jay M. Lipoff.

Crocodile Tooth. Courtesy of M. Konecnik.
(Aquanutz Scuba Diving Charters).

Deer Antler.
Courtesy of Blair Morrow.
(Meg Goddess Designs, Aquanutz).

Deer Teeth in Jawbone.
Courtesy of Michael Tyler Staab (I Hunt Dead Things).

Dire Wolf Tooth (inset). Courtesy of Rick Foresteire.
Dire Wolf Canine. Courtesy of Michael Tyler Staab.
(I Hunt Dead Things).

Drumfish Mouth Plates.
Courtesy of Erin Osborne.
(Charleston Center for Paleontology).

Glyptodont Scute.
Courtesy of Michael Konecnik (Aquanutz).

Glyptodont. Florida Museum of Natural History.
Dr. Jay M. Lipoff.

Gomphothere Tooth. Specimen #CCP-258. Courtesy of the Charleston Center for Paleontology.

Ground Sloth Tooth.
Courtesy of Michael Konecnik (Aquanutz).

Ground Sloth Claw.
Courtesy of Michael Tyler Staab (I Hunt Dead Things).

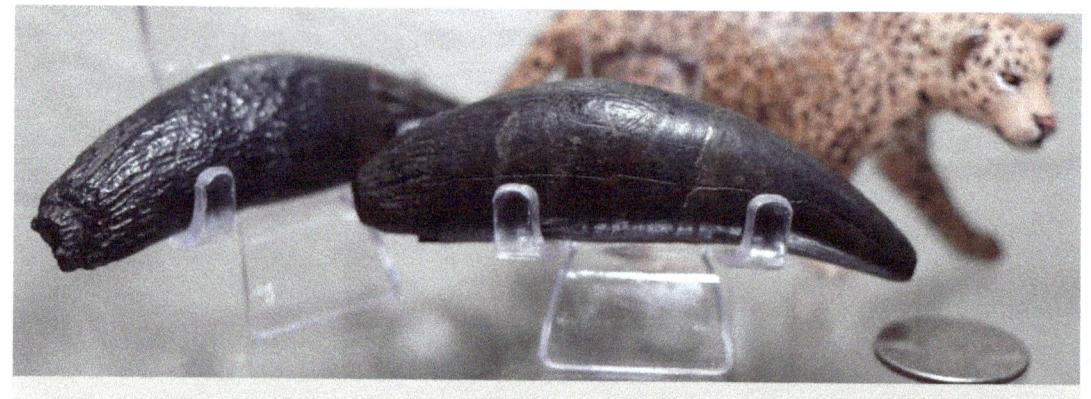

Jaguar Tooth. Courtesy of Michael Tyler Staab. (I Hunt Dead Things).

Dolphin & 2 Sperm Whale Teeth.
Courtesy of Michael Konecnik (Aquanutz).

Two Whale Ear Ossicles and a Dolphin Ear Bone.
Dr. Jay M. Lipoff.

Walrus Tusk. Courtesy of Michael Tyler Staab.
(I Hunt Dead Things).

Variety of Whale Teeth. All from one trip.
Courtesy of Aquanutz Scuba Diving Charters.

Indian Jewelry made from Various Species. Courtesy of Daniel Reed.

Meg Teeth made into Jewelry. Courtesy of Gail and Zack Richardson.

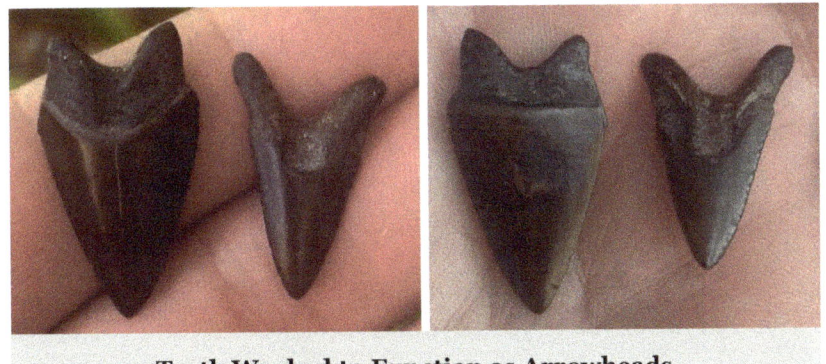

Teeth Worked to Function as Arrowheads.
Courtesy of Gail and Zack Richardson.

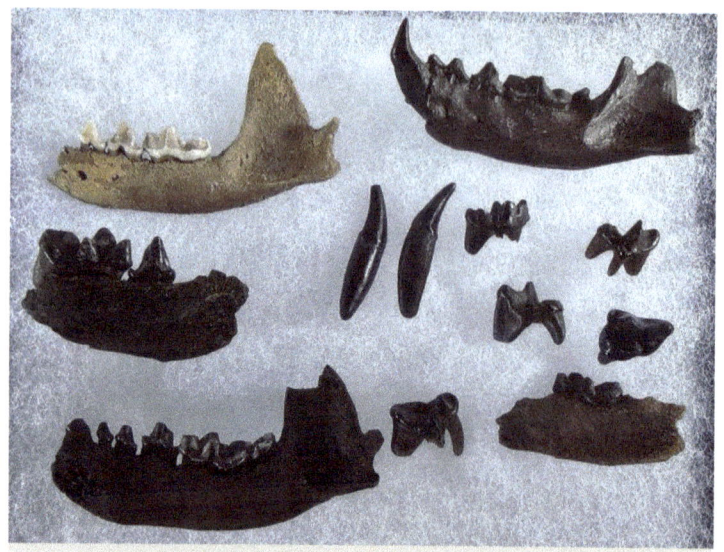

Lontra canadensis, N. American River Otter.
Courtesy of Josh Galloway.

Ray Plates and Barbs. Dr. Jay M. Lipoff.

Squalodon Molar Tooth. Courtesy of Becca LaBrie.

Squalodon Incisor Tooth.
Courtesy of Danny Case.

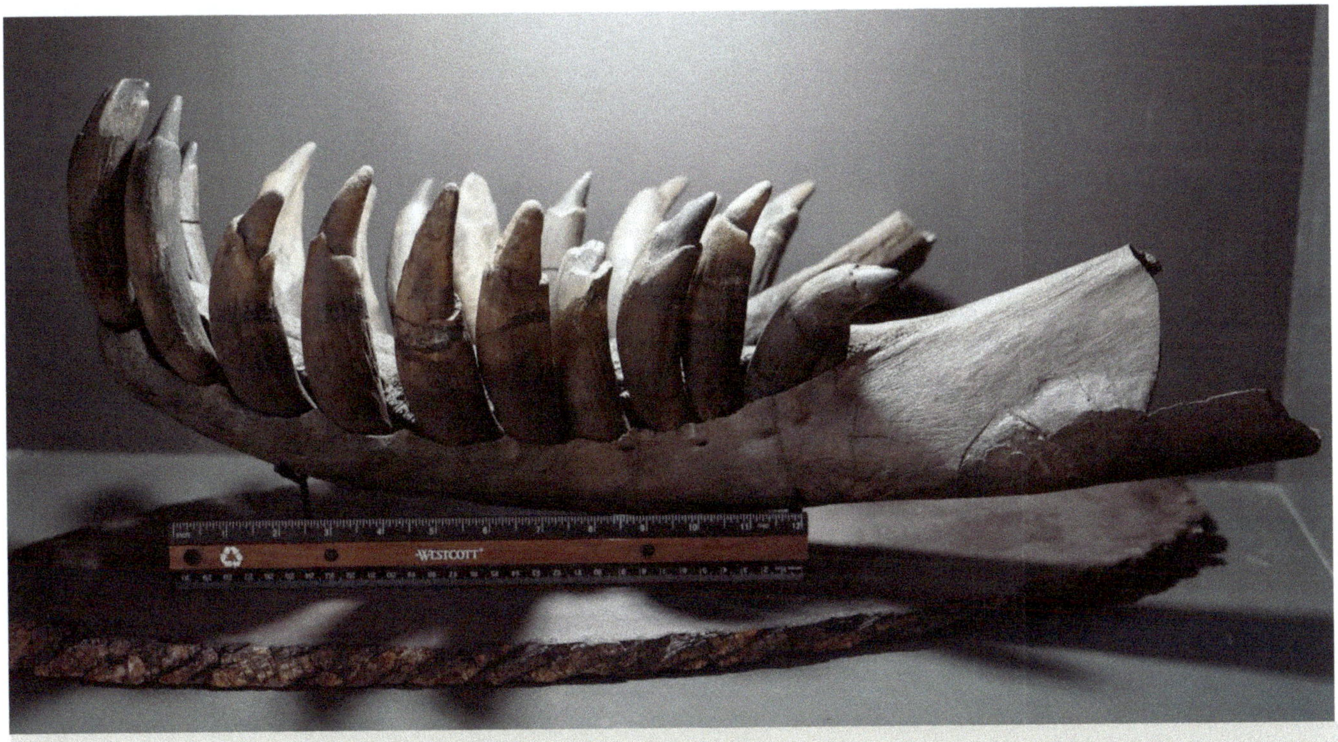

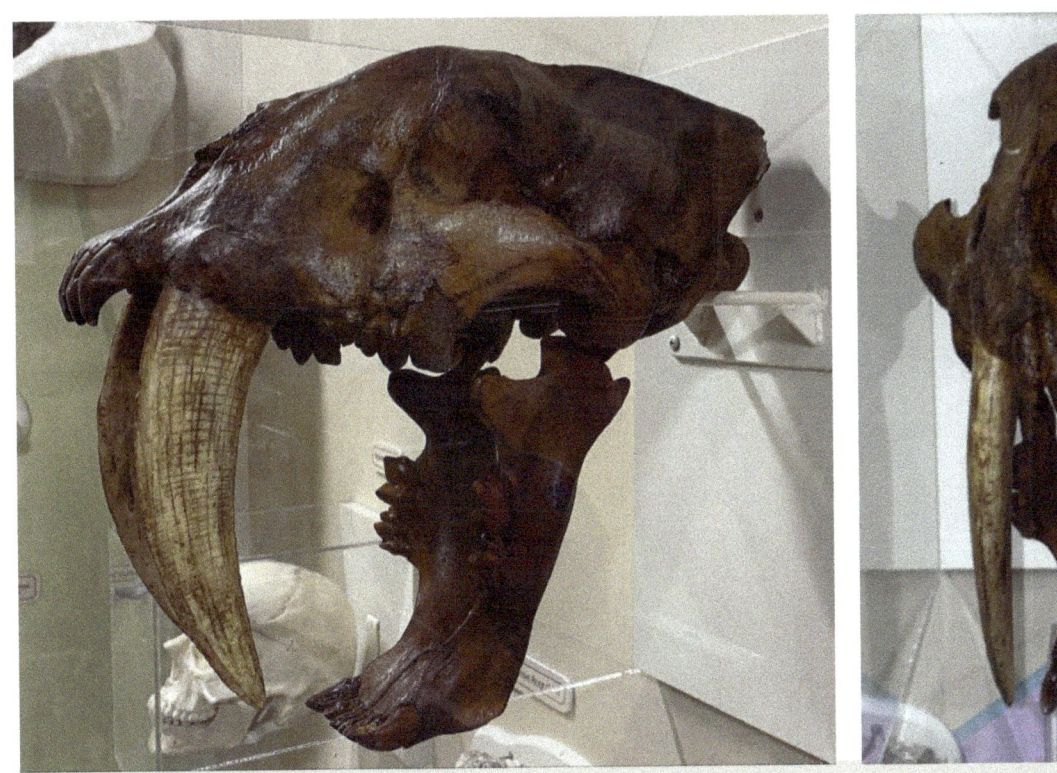

A Small Sperm Whale Under 15 Feet (~4.6 m) Long, *Kogiopsis floridana.*
Courtesy of Michael Tyler Staab (I Hunt Dead Things).

Smilodon or Saber Tooth Tiger. Calvert Marine Museum, Solomons Island, MD. Dr. Jay M. Lipoff.

Tapir Teeth in Jawbone and Individual Teeth.
Courtesy of Michael Tyler Staab (I Hunt Dead Things).

Turtle and Tortoise Shell Fragments. Dr. Jay M. Lipoff.

World War II .50 Caliber Shell Casings.
Dr. Jay M. Lipoff.

World War II .50 Caliber Shell Casings.
Dr. Jay M. Lipoff.

WWII .50 Caliber Shell Casing and Two Projectiles (small) from Practice Rounds.
Courtesy of Michael Konecnik (Aquanutz Scuba Diving Charters).

Variety Pack of Fossils from the St. Mary's County, MD Region.
Dr. Jay M. Lipoff.

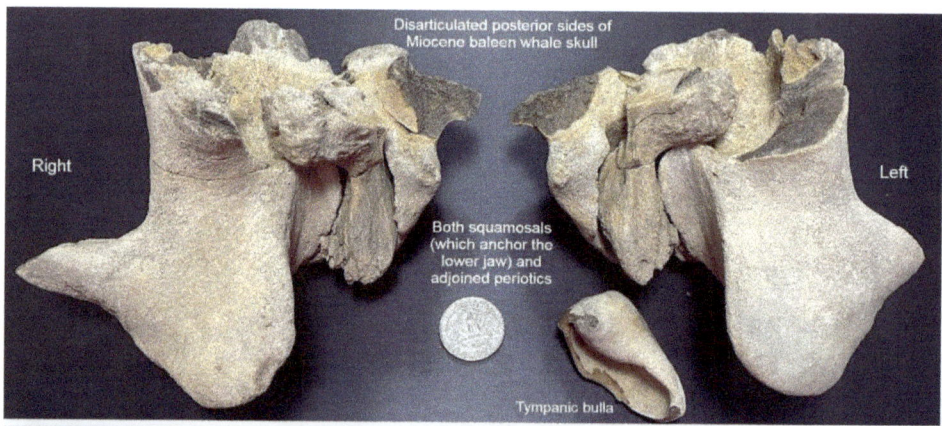

Disarticulated posterior sides of
Miocene baleen whale skull

Right

Left

Both squamosals
(which anchor the
lower jaw) and
adjoined periotics

Tympanic bulla

Anterior end, baleen whale lower jaw

Dolphin
humerus

Worn parts of the rostrum
(premaxillae) of dolphins

Anterior end, baleen whale lower jaw

Metatarsal or phalange bones (3)
from dolphin flipper

Dolphin
or small
whale humerus

Whale and Dolphin Fossils from the Potomac River, MD.
Dr. Jay M. Lipoff.

The Distal End of a Femur from a Columbian Mammoth.
Dr. Jay M. Lipoff.

The same femur, all cleaned up. After removing a considerable amount of growth by soaking the fossil in a vinegar and water solution, I used black shoe polish and a propane torch to ensure the color would penetrate the fossil and restore its black coloring. This technique was used on a trilobite by my friend Alan Langheinrich of Lang's Fossils after a day of digging Eurypterid remipes fossils in Ilion, NY.

Bone Valley Hubbell Meg on Botryoidal Fossil Coral. Courtesy of Dr. Adam Bozeman.

The Total Megalodon Tooth Package

While browsing my Facebook fossils feeds and admiring all the incredible finds that make me jealous, one popped up, and I knew I had to get some pictures for you to enjoy. Blindly, I messaged the owner regarding the latest addition to his collection, hoping he would allow me to showcase his tooth. Like many of the amazing fossil hunters who have donated photos to this project, he also replied, "Yes." All these images were taken by and are the courtesy of Jerome Dinh.

They encompass many aspects discussed in this book, such as the large size of Otodus megalodon teeth, their amazing colors, enamel peeling, predation marks, pathological variations, wonderful serrations, and more. This tooth, measuring 4.9 inches (~12.45 cm), believed to be approximately 4 to 7 million years old and features a serrated bite mark from a megalodon. It was found in the Hawthorne Formation in Beaufort County, North Carolina. It is the total package.

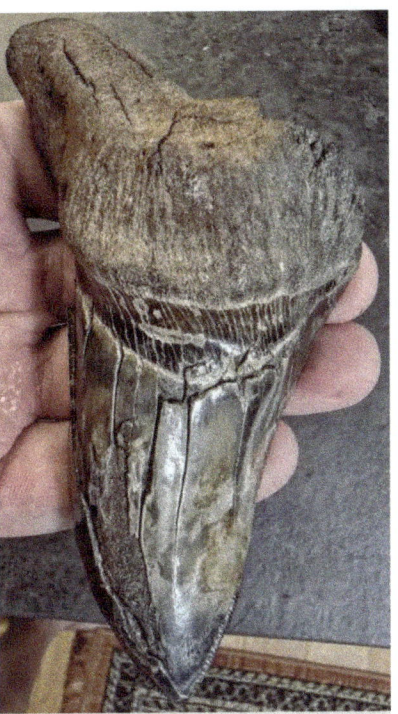

Front (left) **and Back** (right) **Views of the Tooth.**

Left (left) **and Right** (right) **Views of the Tooth.**

Apex View (left) **and Lingual Side of the Tooth with Bite Marks** (right).

Megalodon Bite Marks Near the Tip of the Tooth.

Hippo Fossils from Indonesia.
Courtesy of Ryan Meyer.

Collage of Hemipristis from Bone Valley.
Courtesy of Skylar Vertes.

Laws of Nature. © iStock. Credit: Baris-Ozer.

WEBSITES AND PLACES TO EXPLORE

Museums – Megalodons are all over the world (Online and on Facebook)

American Museum of Natural History – www.amnh.org/exhibitions/sharks

Arizona Museum of Natural History – www.arizonamuseumofnaturalhistory.org

Aurora Fossil Museum – www.aurorafossilmuseum.org

Buena Vista Museum of Natural History & Science – www.buenavistamuseum.org

Calvert Marine Museum – www.calvertmarinemuseum.com

Florida Museum – www.floridamuseum.ufl.edu

Houston Museum of Natural Science – www.hmns.org

Indiana Dinosaur Museum – www.indianadinosaurmuseum.org

Mace Brown Museum of Natural History – charleston.edu/mace-brown-museum/index.php

MOTE Marine Laboratory & Aquarium – www.mote.org

National Aquarium in Baltimore, Maryland – www.aqua.org

North Carolina Museum of Natural Sciences – www.naturalsciences.org

Silver River Museum & Environmental Education Center – www.silverrivermuseum.com

Smithsonian National Museum of Natural History – www.ocean.si.edu

South Carolina State Museum – www.scmuseum.org

Tellus Science Museum – www.tellusmuseum.org

THE BISHOP Museum of Science and Nature – www.bishopscience.org

Virginia Museum of Natural History – www.vmnh.net

Diving and River Fossil Hunting (Eastern U.S.) – (most are on Facebook)

Aquanutz Scuba Dive Charters – www.aquanutz-diving.com

Aristakat Scuba Diving Charter – www.aristakat.com

Black Gold Fossil Charters – www.blackgoldfossilcharters.com

Bone Valley Fossil Farm – www.bonevalleyfossilfarm.com

Burnt & Sandy – burntandsandy.com

Canoe Outpost Peace River – www.facebook.com/CanoeOutpostPeaceRiver

Charleston Fossil Adventures, LLC – www.chsfossiladventures.com

Florida Panhandle Springs and Fossils – www.facebook.com/EconfinaSprings

Florida Program of Vertebrate Paleontology – For a **PERMIT**, (352)-273-1821 vertpaleo@floridamuseum.ufl.edu

Fossil Junkies – www.fossiljunkies.com

Fossil Driven Adventures – fossildrivenadventures.com

Fossil Expeditions – www.fossilexpeditions.com

Fossil Funatics – www.fossilfunatics.com

Fossil Recovery Exploration – www.fossilrecoveryexploration.com

H2O Adventures & More LLC – www.h2oadventuresandmore.com

International Diving Educators Assoc. – www.idea-scuba.com/fossil-hunter

Island Cove Charters – www.islandcovecharters.com

Ladies Fossil Hunter Connection – FL – on Facebook

Megalodon Adventures – www.megalodonadventures.com

Paleo Discoveries – www.paleodiscoveries.com

Palmetto Fossil Excursions – www.palmetto-fossil-excursions.com

Peace Paleo Adventures – www.peacepaleoadventures.com

Peace River Charters – www.peacerivercharters.com

Shark Tooth Island Excursion – www.bullrivermarina.com

SIFT Tours – www.sifttours.com

Top 2 Bottom Offshore Charters – www.top2bottomcharters.com

Venice Dive Center & Charter – www.venicedivecenter.com/charter

Fossil Information and Gifts

Do your own searches, as some sites may have closed. There are too many fossil pages and groups to list. Before purchasing anything, do your research on the item and the seller. It's like buying a diamond; the size, color, whether natural or repaired, condition (including serrations, enamel, wear, etc.), market, competition, and availability all affect the price.

Black River Fossils – www.www.blackriverfossils.org

Black Water Recovery Divers–blackwaterrecoverydivers.com/shop/meg-teeth and on Facebook

Calvert Fossils and Trip Reports – on Facebook

Carolina Shark Tooth Hunters – on Facebook

Charleston Center for Paleontology – charlestoncenterforpaleo.org and on Facebook

Charleston Fossil Adventures, LLC – www.chsfossiladventures.com and on Facebook

Charmouth Fossils (UK) – www.charmouthfossils.com and on Facebook

Coastal Collector by Nautical Necklaces – on Facebook

Code Black Fossils – www.codeblackfossils.com and on Facebook

Dark Water Megs – darkwatermegs.com

Dig Dive Discover – www.youtube.com/digdivediscover and on Facebook

Digging Science – diggingscience.com, YouTube and on Facebook

East Coast Fossil Club – on Facebook

Earth Relics Jewelry Company – www.earthrelics.com and on Facebook

Earth Treasures – www.earthtreasuresfl.com and on Facebook

Every Day I'm Shoveling – on Facebook

FossilEra – www.www.fossilera.com and on Facebook

Fernandina Fossils – www.fernandinafossils.com

Fossils of Calvert Cliffs Maryland – on Facebook

Florida Fossilized – www.floridafossilized.com

Florida Fossil Hunters – www.floridafossilhunters.com and on Facebook

Florida Hunters of Florida – on Facebook

Florida Panhandle Springs and Fossils – on Facebook

Fossil Information & Identification – on Facebook

Florida Shark Tooth Hunters – on Facebook

Fossil Frenzy – www.sharktoothsifter.com

Fossil Guy – www.fossilguy.com

Girls That Scuba – www.girlsthatscuba.com/megalodon-fossil-diving-florida

H2O Adventures & More LLC – www.h2oadventuresandmore.com

I Hunt Dead Things – www.ihdt.store and on Facebook

Internat. Diving Educators Assoc., IDEA – www.idea-scuba.com/fossil-hunter and on Facebook

Lost World Excavations – www.youtube.com/@Lostworldexcavations and on Facebook

Mark Kostich Photography – www.snowleopard1.com

Meg Goddess Designs – www.meggoddessdesigns.com and on Facebook

Megalodon Hunters – on Facebook

Megalodon Maniacs – on Facebook

Megalodon Shark Teeth and Fossils – on Facebook

MegaTeeth Fossils – www.megateeth.com and on Facebook

Megs of Hope – www.megsofhope.org

Nautical Necklaces – on Facebook

Paleontology – on Facebook

Pey Megalodon – on Facebook

Rock Identification Group – on Facebook

Rock Seeker – www.rockseeker.com/where-to-find-megalodon-teeth and on

sea b's creations – www.seabscreations.etsy.com and on Facebook

Seas the Day Charleston – seasthedaycharleston.com

Shark Frenzy / Fossil Frenzy – www.www.sharktoothsifter.com

Shark Life Apparel – www.www.sharklifeapparel.com and on Facebook

Sharksteeth.com – SharksTeeth.com and on Facebook

Shark Tooth Hunters Of The Carolinas – on Facebook

SHRKco – www.shrkco.com and on Facebook

Southwest Florida Shark Tooth Lovers – Buying & Selling Community – on Facebook

Summerville SC Fossil Hunting Edu. – on Facebook

SW FL Fossil and Shark Tooth EDU-Group – on Facebook

The Fossil Forum – on Facebook

The Fossil Exchange – thefossilexchange.com

Unforgettable Oddities – on Facebook

Venice Florida Shark Tooth Hunting Fans (The Original) – on Facebook

Zack and Gail's Fossilized Sharks Tooth Jewelry and More – on Facebook

A Collection of *Otodus chubutensis* on Blue Mountain Jasper. Courtesy of Dr. Adam Bozeman.

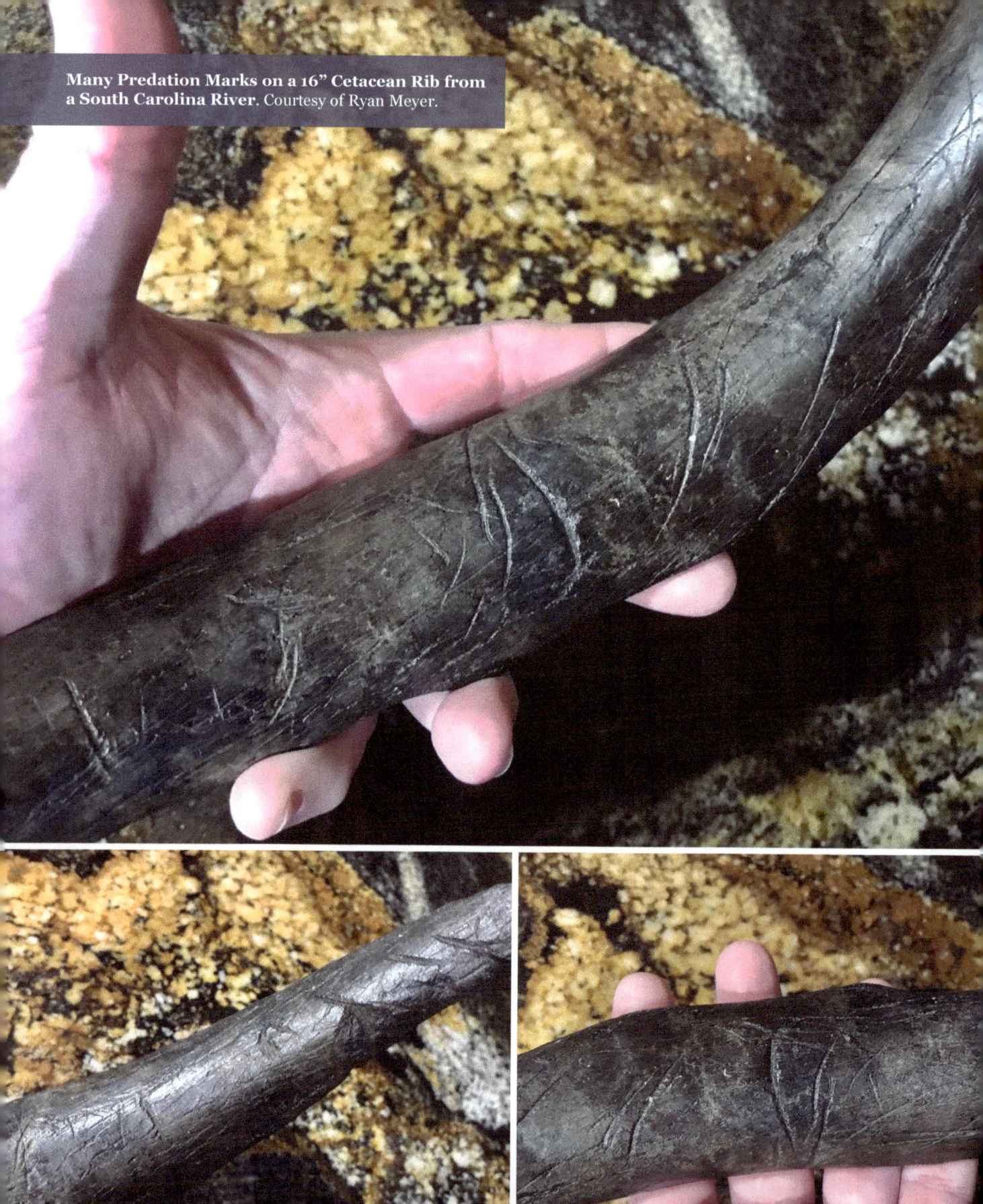

Many Predation Marks on a 16" Cetacean Rib from a South Carolina River. Courtesy of Ryan Meyer.

Dr. Jay M. Lipoff.

Finding the distal end of a Columbian Mammoth off the coast of Venice, Florida, with Aquanutz Scuba Diving Charters, and on the right, 1 of 2 pathological horseshoe crabs found along the Patuxent River in Southern Maryland.

ABOUT THE AUTHOR

Like you, Dr. Jay has always been passionate about fossils and diving. Scuba diving for fossils is a perfect match—it's like an Easter egg hunt for prehistoric treasures. His love for searching for Megalodon teeth, the strong friendships he cultivates during these adventures, and the opportunity to learn more about Megalodon were reasons why he wrote this book.

Although he is a Doctor of Chiropractic, he enjoys writing and playing music, creating art, sports, and teaching, which has led him to write four other books. They are "*Back at Your Best: Balancing the Demands of Life with the Needs of Your Body*", "*Are You Ready To Be A HERO?*", "*Super Coco, Will You Be My Friend?*", and "*Donny the Megalodon and the JAWsome Miocene Adventure.*"

Dr. Jay has been awarded the Arnold Schwarzenegger Legacy Award for promoting health, wellness, and fitness and, most importantly, for inspiring others. He has also received the William Donald Schaefer Award for his commitment to serving those in need and improving their communities.

REFERENCES

Introduction References

1. Murphy, J. & Nance, R. (2004). How Do Supercontinents Assemble? *American Scientist. 92(4)*, 324-333. https://doi.org/10.1511/2004.48.935

2. Le Pichon, X., Sengor, A. M. C., Jellinek, M., Lenardic, A. & İmren, C. (2023). Breakup of Pangea and the Cretaceous Revolution. *Tectonics. 42(2)*, e2022TC007489. https://doi.org/10.1029/2022TC007489

3. Hine, A. C., Dunn, S. C. & Locker, S. D. (2013). *Geologic Beginnings of the Gulf of Mexico with Emphasis on the Formation of the De Soto Canyon.* Deep-C Consortium. August 14, 2013. https://deep-c.coaps.fsu.edu/news-and-multimedia/in-the-news/geologic-beginnings-of-the-gulf-of-mexico-with-emphasis-on-the-formation-of-the-de-soto-canyon

4. Bostick, K., Johnson, S. & Main, M. (2018). Florida's Geologic History. *EDIS. 2018(6)*. https://doi.org/10.32473/edis-uw208-2018

5. Hine, A. C. (2013). *Geologic History of Florida: Major Events that Formed the Sunshine State.* University Press of Florida. June 18, 2013.

6. Gardulski, A. (1991). Evolution of a deep-water carbonate platform: Upper Cretaceous to Pleistocene sedimentary environments on the west Florida margin. Marine Geology, Volume 101, Issues 1–4, Pages 163-179. https://doi.org/10.1016/0025-3227(91)90069-G

7. Perez, V. (2022). The chondrichthyan fossil record of the Florida Platform (Eocene–Pleistocene). *Paleobiology, 48*, 1-33. https://doi.org/10.1017/pab.2021.47

8. Missimer, T. M., & Maliva, R. G. (2017). Late Miocene fluvial sediment transport from the southern Appalachian Mountains to southern Florida: An example of an old mountain belt sediment production surge. *Sedimentology, V. 64*, p. 1846–1870. https://doi.org/10.1111/sed.12377

9. Poag, C. W. (2022). Bolide impact effects on the West Florida Platform, Gulf of Mexico: End Cretaceous and late Eocene. *Geosphere, 18 (3)*, 1077–1103. doi: https://doi.org/10.1130/GES02472.1

10. Katz, M., Miller, K., Wright, J. et al. (2008). Stepwise Transition from the Eocene Greenhouse to the Oligocene Icehouse. *Nature Geoscience,1*, 329–334. https://doi.org/10.1038/ngeo179

11. Petuch, E.J., & Roberts, C. (2007). *The Geology of the Everglades and Adjacent Areas (1st ed.).* CRC Press. https://doi.org/10.1201/9781420045598

12. Huebscher, C., et al. (2023). Reading the sediment archive of the Eastern Campeche Bank (southern Gulf of Mexico): from the aftermath of the Chicxulub impact to Loop Current variability. *Marine Geophysical Research, 44(2)*, https://doi.org/10.1007/s11001-023-09514-3

13. Bryant, J. D., McFadden, B. J. & Mueller, P. A. (1992). Improved chronologic resolution of the Hawthorn and the Alum Bluff Groups in northern Florida: Implications for Miocene chronostratigraphy. *GSA Bulletin 1992, 104 (2)*, 208–218. https://doi.org/10.1130/0016-7606(1992)104<0208:ICROTH>2.3.CO;2

14. Screaton, E., Martin, J. B., Ginn, B. & Smith, L. (2004). *Conduit properties and karstification in the unconfined Floridan Aquifer"* Ground Water v. 42, p. 338–346. https://doi.org/10.1111/j.1745-6584.2004.tb02682.x

15. White, W. B. (2002). Karst hydrology: recent developments and open questions. *Engineering Geology, v. 65*, p. 85–105. https://doi.org/10.1016/S0013-7952(01)00116-8

16. Pimiento, C., Griffin, J.N., Clements, C.F. et al. (2017). The Pliocene marine megafauna extinction and its impact on functional diversity. *Nature Ecology & Evolution, 1*, 1100–1106. https://doi.org/10.1038/s41559-017-0223-6

17. Ji, W., Robel, A., Tziperman, E., & Yang, J. (2021). Laurentide ice saddle mergers drive rapid sea level drops during glaciations. *Geophysical Research Letters, 48*, e2021GL094263. https://doi.org/10.1029/2021GL094263

Chapter 1. Classification of Fish References

1. Ng, C., Ooi, P., Wong, W-L. & Khoo, G. (2017). A Review of Fish Taxonomy Conventions and Species Identification Techniques. *Journal of Survey in Fisheries Sciences, 4(1)*, 54-93. https://doi.org/10.18331/SFS2017.4.1.6

2. Gao, G., Sun, Z., Mu, G., Yin, H., & Ren, Y. (2024). Research on marine fish classification and recognition based on an optimized ResNet50 model. *Marine and Coastal Fisheries: Dynamics, Management, and Ecosystem Science, 16*, e10317. https://doi.org/10.1002/mcf2.10317

3. Brazeau, M. D., & Friedman, M. (2015). The origin and early phylogenetic history of jawed vertebrates. *Nature, 520(7548)*, 490–497. https://doi.org/10.1038/nature14438

4. Anderson, P., Friedman, M., Brazeau, M. et al. Initial radiation of jaws demonstrated stability despite faunal and environmental change. *Nature, 476*, 206–209 (2011). https://doi.org/10.1038/nature10207

5. Lund-Ricard, Y. & Boutet, A. (2021). *Current Trends in Chondrichthyes*. Chapter 23. https://doi.org/10.1201/9781003217503-23

6. Zhu, M., Zhao, W., Jia, L., et al. (2009). The oldest articulated osteichthyan reveals mosaic gnathostome characters. *Nature, 458*, 469–474 https://doi.org/10.1038/nature07855

7. Dean, M. N., et al. (2024). *Cartilaginous fish skeletal tissues, Encyclopedia of Fish Physiology (Second Edition)*. Academic Press, Pages 452-459, https://doi.org/10.1016/B978-0-323-90801-6.00036-7

8. Serena, F., et al. (2020). Species diversity, taxonomy and distribution of Chondrichthyes in the Mediterranean and Black Sea. *The European Zoological Journal, 87:1*, 497-536. https://doi.org/10.1080/24750263.2020.1805518

Chapter 1. A Brief History of Sharks References

1. Boessenecker, R. W., Ehret, D. J., Long, D. J., Churchill, M., Martin, E, & Boessenecker, S. J. (2019). The Early Pliocene extinction of the mega-toothed shark Otodus megalodon: a view from the eastern North Pacific. *Peer J*, 7:e6088. https://doi.org/10.7717/peerj.6088

2. The PLOS ONE Staff (2015) Correction: When Did Carcharocles megalodon Become Extinct? A New Analysis of the Fossil Record. PLoS ONE, 10(1): e0117877. https://doi.org/10.1371/journal.pone.0117877

3. Pimiento, C. & Balk, M. A. (2015). Body-size trends of the extinct giant shark Carcharocles megalodon: a deep-time perspective on marine apex predators. Paleobiology, 41(3), 479-490. https://doi.org/10.1017/pab.2015.16

4. Shimada. K., Bonnan, M. F., Becker, M. A. & Griffiths, M. L. (2021). Ontogenetic growth pattern of the extinct megatooth shark Otodus megalodon - implications for its reproductive biology, development, and life expectancy. *Historical Biology, An International Journal of Paleobiology, Volume 33, Issue 12*, Pages 3254-3259. https://doi.org/10.1080/08912963.2020.1861608

Chapter 1. Survival Instincts References

1. Strickland, B.A., Massie, J.A., Viadero, N. et al. (2020). Movements of Juvenile Bull Sharks in Response to a Major Hurricane Within a Tropical Estuarine Nursery Area. *Estuaries and Coasts, 43*, 1144–1157. https://doi.org/10.1007/s12237-019-00600-7

2. Heupel, M.R., Simpfendorfer, C.A. & Hueter, R.E. (2003), Running before the storm: blacktip sharks respond to falling barometric pressure associated with Tropical Storm Gabrielle. Journal of Fish Biology, 63: 1357-1363. https://doi.org/10.1046/j.1095-8649.2003.00250.x

3. Learn, J. R. (2019). *Where Do Sharks Go in Hurricanes?* Hakai Magazine, Coastal Science and Societies. December 18, 2019. https://hakaimagazine.com/news/where-do-sharks-go-in-hurricanes/

4. Gutowsky, L. F. G., et al. (2021). Large sharks exhibit varying behavioral responses to major hurricanes. *Estuarine, Coastal and Shelf Science, Volume 256*, 107373, https://doi.org/10.1016/j.ecss.2021.107373

5. Shiffman, D. (2018). *Sharks*. Ocean, Smithsonian Institute. April 2018. https://ocean.si.edu/ocean-life/sharks-rays/sharks

Chapter 2. Fossil Clues References

1. Boessenecker, R. W., Ehret, D. J., Long, D. J., Churchill, M., Martin, E, & Boessenecker, S. J. (2019). The Early Pliocene extinction of the mega-toothed shark Otodus megalodon: a view from the eastern North Pacific. *Peer J*, 7:e6088. https://doi.org/10.7717/peerj.6088

2. Spurgeon, E. et al. (2024). The influence of micro-scale thermal habitat on the movements of juvenile white sharks in their Southern California aggregation sites. *Frontiers In Marine Science, Vol 11.* https://doi.org/10.3389/fmars.2024.1290769

3. Frontiers in Marine Science. (2024). *Baby white sharks prefer being closer to shore, scientists find.* PHYS.org. April 19, 2024. https://phys.org/news/2024-04-baby-white-sharks-closer-shore.html

4. Shimada, K., Bonnan, M. F., Becker, M. A., & Griffiths, M. L. (2021). Ontogenetic growth pattern of the extinct megatooth shark Otodus megalodon—implications for its reproductive biology, development, and life expectancy. *Historical Biology, 33(12),* 3254–3259. https://doi.org/10.1080/08912963.2020.1861608

5. Taylor & Francis Group. (2021, January 10). *Megalodons gave birth to large newborns that likely grew by eating unhatched eggs in the womb.* ScienceDaily. Retrieved June 25, 2024. from www.sciencedaily.com/releases/2021/01/210110192433.htm

6. Ehret, D. J., Hubbell, G., & Macfadden, B. J. (2009). Exceptional preservation of the white shark Carcharodon (Lamniformes, Lamnidae) from the early Pliocene of Peru. *Journal of Vertebrate Paleontology, 29(1),* 1–13. https://doi.org/10.1671/039.029.0113

Chapter 2. Growing Up Big References

1. Shimada. K., Bonnan, M. F., Becker, M. A. & Griffiths, M. L. (2021). Ontogenetic growth pattern of the extinct megatooth shark Otodus megalodon - implications for its reproductive biology, development, and life expectancy. *Historical Biology, An International Journal of Paleobiology, Volume 33, Issue 12*, Pages 3254-3259. https://doi.org/10.1080/08912963.2020.1861608

2. Herraiz J. L., Ribé J., Botella H., Martínez-Pérez, C. & Ferrón H. G. (2020). Use of nursery areas by the extinct megatooth shark Otodus megalodon (Chondrichthyes: Lamniformes). *Biology Letters, Volume 16, Issue 11.* Pp. 1-7. http://doi.org/10.1098/rsbl.2020.0746

3. Griffiths, M., Eagle, R., Kim, S., Flores, R., Becker, M., Maisch IV, H., Trayler, R., Chan, R., McCormack, J., Akhtar, A., Tripati, A. & Shimada, K. (2023). Endothermic physiology of extinct megatooth sharks. *PNAS, Vol 120, No 27,* e2218153120. https://doi.org/10.1073/pnas.2218153120

4. Sato, K., Nakumura, M., Tomita, T., Toda, M., Miyamot. J. & Nozu, R. (2016). How great white sharks nourish their embryos to a large size: evidence of lipid histotrophy in lamnoid shark reproduction. *Biol Open, 5 (9),* 1211–1215. https://doi.org/10.1242/bio.017939

5. Southeast. (2024). *Necropsy Offers Rare Opportunity to Study White Shark Biology.* NOAA Fisheries. February 28, 2024. https://www.fisheries.noaa.gov/feature-story/necropsy-offers-rare-opportunity-study-white-shark-biology

6. Miller, E., Wails, C.N. & Sulikowski, J. (2022). It's a shark-eat-shark world, but does that make for bigger pups? A comparison between oophagous and non-oophagous viviparous sharks. *Rev Fish Biol Fisheries, 32,* 1019–1033. https://doi.org/10.1007/s11160-022-09707-w

7.   Denoel, M. & Demars, B. (2008). The benefits of heterospecific oophagy in a top predator. *Acta Oecologica. 34.* 74-79. https://doi.org/10.1016/j.actao.2008.03.004

8.   Weisberger, M. (2021*). Megalodon shark mamas had human-size cannibal babies.* Live Science. January 11, 2021. https://www.livescience.com/megalodon-babies-human-size. html#:~:text=Its%20young%20were%20the%20largest%20live%20babies%20in%20the%20 shark%20family.&text=Megalodon%20was%20the%20biggest%20predatory,as%20the%20 average%20basketball%20player.

9.   Wibowo, A., Basukriadi, A., Nurdin, E. & Mubarok, M. (2021). Habitat Preference Modeling of Prehistoric Giant Shark Megalodon During Miocene in Bentang Formation of West Java Coast. *Jurnal Biodjati. 6(2),* 264-272. https://doi.org/10.15575/biodjati.v6i2.14115

10.   Cooper, J.A., Pimiento, C., Ferrón, H.G. et al. (2020). Body dimensions of the extinct giant shark Otodus megalodon: a 2D reconstruction. *Sci Rep, 10,* 14596 https://doi.org/10.1038/ s41598-020-71387-y

11.   Cooper, J. A. et al. (2022). *The extinct shark Otodus megalodon was a transoceanic superpredator: Inferences from 3D modeling.* Science Advances. Published online August 17, 2022. https://doi.org/10.1126/sciadv.abm9424

12.   Shimada, K. et al. (2025). Reassessment of the possible size, form, weight, cruising speed, and growth parameters of the extinct megatooth shark, Otodus megalodon (Lamniformes: Otodontidae), and new evolutionary insights into its gigantism, life history strategies, ecology, and extinction. *Palaeontologia Electronica, 28(1),* a12. https://doi.org/10.26879/1502

13.   Bennett, M. (2024). *The Megalodon's Forceful Bite Was 10 Times More Powerful Than a Great White Shark's.* Discover Magazine. February 13, 2024. https://www.discovermagazine. com/the-sciences/the-megalodons-forceful-bite-was-10-times-more-powerful-than-a-great-white

14.   Edmonds H. M. & Glowacka H. (2020). The ontogeny of maximum bite force in humans. *J. Anatomy, 237(3),* 529–542. https://doi.org/10.1111/joa.13218

15.   Rayne, E. (2023). *Which animals have the strongest bite?* Live Science. March 19, 2023. https://www.livescience.com/which-animals-have-the-strongest-bite

16.   Perez, V. L., Leder, R. M. & Badaut, T. (2021). Body length estimation of Neogene macrophagous lamniform sharks (Carcharodon and Otodus) derived from associated fossil dentitions. *Palaeologica Electronica. Article number: 24.1.a09.* https://doi. org/10.26879/1140

17.   Burls, N. J., Bradshaw, C. D., De Boer, A. M., Herold, N., Huber, M., et al. (2021). Simulating Miocene warmth: Insights from an opportunistic multi-model ensemble (MioMIP1). *Paleoceanography and Paleoclimatology, 36,* e2020PA004054. https://doi. org/10.1029/2020PA004054

18.   Sosdian, S., Coxall, H., Steinthorsdottir, M. & Lawrence, K. (2024) Miocene temperature portal. Dataset version 3. *Bolin Centre Database.* https://doi.org/10.17043/miocene-temperature-portal-3

19. Godfrey, S. J., Ellwood, M., Groff, S. & Verdin, M. S. (2018). Carcharocles-bitten odontocete caudal vertebrae from the Coastal Eastern United States. *Acta Palaeontologica Polonica, 63 (3)*, 463-468. https://doi.org/10.4202/app.00495.2018

20. Krisch, J. A. (2022). *Terrifying megalodon attack on whale revealed in 15-million-year-old fossils*. Live Science. September 9, 2022. https://www.livescience.com/megalodon-vs-whale-failed-predation

21. Godfrey, Stephen & Nance, John & Riker, Norman. (2021). Carcharocles-bitten sperm whale tooth from the Neogene of the Coastal Eastern United States. *Acta Palaeontologica Polonica, 66(3)*, https://doi.org/10.4202/app.00820.2020

Chapter 3. Coastal Nurseries References

1. Shimada, K., Maisch, H. M., Perez, V. J., Becker, M. A., & Griffiths, M. L. (2022). Revisiting body size trends and nursery areas of the Neogene megatooth shark, Otodus megalodon (Lamniformes: Otodontidae), reveals Bergmann's rule possibly enhanced its gigantism in cooler waters. *Historical Biology, 35*(2), 208–217. https://doi.org/10.1080/08912963.2022.2032024

2. Villafaña, J.A., Hernandez, S., Alvarado, A. et al. (2020). *First evidence of a palaeo-nursery area of the great white shark*. Scientific Reports, 10, Article 8502. https://doi.org/10.1038/s41598-020-65101-1

3. Landini, W., Collareta, A., Pesci, F., Di Celma, C., Urbina Schmitt, M. & Bianucci, G. (2017). A secondary nursery area for the copper shark Carcharhinus brachyurus from the late Miocene of Peru. *Journal of South American Earth Sciences. 78.* https://doi.org/0.1016/j.jsames.2017.07.003

4. Miller, A., Gibson, M. & Boessenecker, R. (2021). A megatoothed shark (Carcharocles angustidens) nursery in the Oligocene Charleston Embayment, South Carolina, USA. *Palaeontologia Electronica, 24.* https://doi.org/10.26879/1148

5. Pimiento, C., Ehret, D. J., MacFadden, B. J. & Hubbell, G. (2010). Ancient Nursery Area for the Extinct Giant Shark Megalodon from the Miocene of Panama. *PLoS ONE 5(5)*, e10552. https://doi.org/10.1371/journal.pone.0010552

6. Sato, K., Nakumura, M., Tomita, T., Toda, M., Miyamot. J. & Nozu, R. (2016). How great white sharks nourish their embryos to a large size: evidence of lipid histotrophy in lamnoid shark reproduction. *Biol Open, 5 (9)*, 1211–1215. https://doi.org/10.1242/bio.017939

7. Haiken, M. (2024). *Baby great white shark reveals huge nursery near NYC in scientific first*. National Geographic. June 24, 2024. https://www.nationalgeographic.com/animals/article/babies-great-white-sharks-camera-new-york

8. Ferrón, H. G. (2017). Regional endothermy as a trigger for gigantism in some extinct macropredatory sharks. *PLoS ONE, 12(9)*, e0185185. https://doi.org/10.1371/journal.pone.0185185

9. BBC. (2024). *Why young great white sharks may prefer shallow waters.* bbc.co.uk. April 22, 2024. https://www.bbc.co.uk/newsround/68858008.amp

10. Frontiers in Marine Science. (2024). *Baby white sharks prefer being closer to shore, scientists find.* PHYS.org. April 19, 2024. https://phys.org/news/2024-04-baby-white-sharks-closer-shore.html

11. Tsai, C. (2017). A Miocene breeding ground of an extinct baleen whale (Cetacea: Mysticeti) *PeerJ, 5*:e3711. https://doi.org/10.7717/peerj.3711

12. Miro, J. (2018). *In the Galapagos, an idyllic hammerhead shark nursery.* PHYS.org. January 29, 2018. https://phys.org/news/2018-01-galapagos-idyllic-hammerhead-shark-nursery.html

13. Ashworth, J. (2022). *Megalodon sharks grew biggest in colder waters.* Natural History Museum. March 7, 2022. https://www.nhm.ac.uk/discover/news/2022/march/megalodon-sharks-grew-biggest-colder-waters.html

Chapter 4. Megalodon Taxonomy References

1. Trif, N., Ciobanu, R. & Vlad, C. (2016). The first record of the giant shark Otodus megalodon (Agassiz, 1835) from Romania. *Brukenthal Acta Musei. XI.* 507-526. https://www.researchgate.net/publication/309615167_The_first_record_of_the_giant_shark_Otodus_megalodon_Agassiz_1835_from_Romania

2. Carroll, S. B. (2009). *In a Shark's Tooth, a New Family Tree.* The New York Times. September 14, 2009. https://www.nytimes.com/2009/09/15/science/15creature.html

3. Starr, J. D. (2024). *Louis Agassiz.* Encyclopedia Britannica. June 18, 2024. https://www.britannica.com/biography/Louis-Agassiz

4. Buehler, J. (2024). *Megalodon, the largest shark ever, may have been a long, slender giant.* Science News. January 21, 2024. https://www.sciencenews.org/article/megalodon-largest-shark-fossil-long-body

5. Black, R. (2012). *Great White Shark Ancestry Swims Into Focus.* National Geographic. November 15, 2012. https://www.nationalgeographic.com/science/article/great-white-shark-ancestry-swims-into-focus

6. Long, D. & Waggoner, B. (1996). *Evolutionary Relationships of the White Shark: A Phylogeny of Lamniform Sharks Based on Dental Morphology.* Great White Sharks: The Biology of Carcharodon carcharias. (pp.37-47), Chapter: 5. Academic Press. https://doi.org/10.1016/B978-012415031-7/50006-9

7. Pimiento, C., Cantalapiedra, J. L., Shimada, K., Field, D. J., & Smaers, J. B. (2019). Evolutionary pathways toward gigantism in sharks and rays. *Evolution, 73 (2)*, 588–599. https://doi.org/10.1111/evo.13680

8. Nyberg, K. G., Ciampaglio, C. N., & Wray, G. A. (2006). Tracing the ancestry of the great white shark, Carcharodon carcharias, using morphometric analyses of fossil teeth. *Journal of Vertebrate Paleontology, 26(4)*, 806-814. https://doi.org/10.1671/0272-4634(2006)26[806:ttaotg]2.0.co;2

9.   Brignon, A. (2021). Historical and nomenclatural remarks on some megatoothed shark teeth (Elasmobranchii, Otodontidae) from the Cenozoic of New Jersey (U.S.A.). *Rivista Italiana di Paleontologia e Stratigrafia. Vol. 127, No. 3*, 595-625. https://doi.org/10.13130/2039-4942/16440

10.   Ebersole, J. A. & Ehret, D. J. (2018). A new species of Cretalamna sensu stricto (Lamniformes, Otodontidae) from the Late Cretaceous (Santonian-Campanian) of Alabama, USA. *PeerJ 6*, e4229. https://doi.org/10.7717/peerj.4229

11.   Bazzi, M., Campione, N., Ahlberg, P., Blom, H. & Kear, B. (2021). Tooth morphology elucidates shark evolution across the end-Cretaceous mass extinction. *PLoS Biology. 19(8)*. https://doi.org/10.1371/journal.pbio.3001108

12.   Ballell, A. & Ferrón, H.G. (2021). Biomechanical insights into the dentition of megatooth sharks (Lamniformes: Otodontidae). *Scientific Reports, 11, Article 1232*. https://doi.org/10.1038/s41598-020-80323-z

13.   Shimada, K., Chandler, R. E., Lam, O. L. T., Tanaka, T., & Ward, D. J. (2016). A new elusive otodontid shark (Lamniformes: Otodontidae) from the lower Miocene, and comments on the taxonomy of otodontid genera, including the 'megatoothed' clade. *Historical Biology, 29(5)*, 704–714. https://doi.org/10.1080/08912963.2016.1236795

14.   Ehret, D., MacFadden, B., Jones, D., Devries, T., Foster, D. & Salas-Gismondi, R. (2012). Origin of the White Shark Carcharodon (Lamniformes: Lamnidae) based on recalibration of the Upper Neogene Pisco Formation of Peru. *Palaeontology. 55(6)*. https://doi.org/10.1111/j.1475-4983.2012.01201.x

15.   Diedrich, C. G. (2013). White and megatooth shark evolution and predation origin onto seals, sirenians, and whales. *Natural Science, 5(11)*, 1203-1218. http://dx.doi.org/10.4236/ns.2013.511148

16.   Ferrón, H. G., Martínez-Pérez, C., & Botella, H. (2017). The evolution of gigantism in active marine predators. *Historical Biology, 30(5)*, 712–716. https://doi.org/10.1080/08912963.2017.1319829

17.   Perez, V. J., Godfrey, S. J., Kent, B. W., Weems, R. E., & Nance, J. R. (2018). The transition between Carcharocles chubutensis and Carcharocles megalodon (Otodontidae, Chondrichthyes): lateral cusplet loss through time. *Journal of Vertebrate Paleontology, 38(6)*. https://doi.org/10.1080/02724634.2018.1546732

18.   Sternes, P. C., et al. (2024). *White shark comparison reveals a slender body for the extinct megatooth shark, Otodus megalodon (Lamniformes: Otodontidae)*. Palaeontologia Electronica, Article number: 27.1.a7. https://doi.org/10.26879/1345

19.   Shimada, K. (2019). The size of the megatooth shark, Otodus megalodon (Lamniformes: Otodontidae), revisited. *Historical Biology, 33*, 904 - 911. https://doi.org/10.1080/08912963.2019.1666840

20.   Dolton, H. R., Jackson, A. L., Deaville, R., Hall, J., Hall, G., McManus G., Perkins, M. R., Rolfe, R. A., Snelling, E. P., Houghton, J. D. R., Sims, D. W., & Payne, N. L. (2023a). Regionally endothermic traits in the planktivorous basking sharks Cetorhinus maximus. Endangered Species Research, 51:227–232. https://doi.org/10.3354/esr01257

21. Hobson, M. (2024). *Megalodon didn't look like a 50-foot great white shark, a controversial study claims*. Live Science. January 21, 2024. https://www.livescience.com/animals/extinct-species/controversial-study-claims-megalodon-didnt-look-like-a-50-foot-giant-great-white-shark

22. Shimada, K. et al. (2025). Reassessment of the possible size, form, weight, cruising speed, and growth parameters of the extinct megatooth shark, Otodus megalodon (Lamniformes: Otodontidae), and new evolutionary insights into its gigantism, life history strategies, ecology, and extinction. Palaeontologia Electronica, 28(1):a12. https://doi.org/10.26879/1502

23. Bendix-Almgreen, S. E.: Carcharodon megalodon from the Upper Miocene of Denmark, with comments on elasmobranch tooth enameloid: coronoin. *Bull. geol. Soc. Denmark, vol 32*, pp. 1-32, Copenhagen, November, 15th, 1983. https://doi.org/10.37570/bgsd-1983-32-01

24. Vullo, R. et al. (2024). Exceptionally preserved shark fossils from Mexico elucidate the long-standing enigma of the Cretaceous elasmobranch Ptychodus. *Proceedings of the Royal Society B: Biological Sciences*. 291. 20240262. https://doi.org/10.1098/rspb.2024.0262

25. Vullo, R. et al. (2021). Manta-like planktivorous sharks in Late Cretaceous oceans. *Science, 371*, 1253-1256. https://doi.org/10.1126/science.abc1490

Chapter 5. Apex Predator References

1. Lambert, O., Bianucci, G. & de Muizon, C. (2016). Macroraptorial sperm whales (Cetacea, Odontoceti, Physeteroidea) from the Miocene of Peru. *Zoological Journal of the Linnean Society. 179(2)*. https://doi.org/10.1111/zoj.12456

2. Jeffrey, A. (2016). *Giant killer sperm whales once cruised Australia's waters (and we have a massive tooth to prove it)*. Earth Touch News Network. April 22, 2016. https://www.earthtouchnews.com/discoveries/fossils/giant-killer-sperm-whales-once-cruised-australias-waters-and-we-have-a-massive-tooth-to-prove-it/

3. Lambert, O., Bianucci, G., Post, et al. (2010). The giant bite of a new raptorial sperm whale from the Miocene epoch of Peru. *Nature. 466(7302)*, 105-108. https://doi.org/10.1038/nature09067

4. Fang, J. (2010). Call me Leviathan melvillei. *Nature*. https://doi.org/10.1038/news.2010.322

5. Godfrey, S. J., Nance, J. R. & Riker, N. L. (2021). Otodus-bitten sperm whale tooth from the Neogene of the Coastal Eastern United States. *Acta Palaeontol. Pol., 66(3)*, 599–603. https://www.app.pan.pl/archive/published/app66/app008202020.pdf

6. Godfrey, S. J., Ellwood, M., Groff, S. & Verdin, M. S. (2018). Carcharocles-bitten odontocete caudal vertebrae from the Coastal Eastern United States. *Acta Palaeontologica Polonica, 63 (3)*, 463-468. https://doi.org/10.4202/app.00495.2018

7. Collareta, A., Lambert, O., Landini, W., et al. (2017). Did the giant extinct shark Carcharocles megalodon target small prey? Bite marks on marine mammal remains from the late Miocene of Peru. *Palaeogeography, Palaeoclimatology, Palaeoecology, Volume 469*, Pages 84-91. https://doi.org/10.1016/j.palaeo.2017.01.001

8.   Boessenecker, R. W., Churchill, M., Buchholtz, E. A., Beatty, B. L. & Geisler, J. H. (2020). Convergent Evolution of Swimming Adaptations in Modern Whales Revealed by a Large Macrophagous Dolphin from the Oligocene of South Carolina. *Current Biology, Volume 30, Issue 16*, Pages 3267-3273.e2. https://doi.org/10.1016/j.cub.2020.06.012

Chapter 5. Speed References

1.   Shimada, K., Yamaoka, Y., Kurihara, Y., Takakuwa, Y., Maisch, H. M., Becker, M. A., Eagle, R. & Griffiths, M. L. (2023). Tessellated calcified cartilage and placoid scales of the Neogene megatooth shark, Otodus megalodon (Lamniformes: Otodontidae), offer new insights into its biology and the evolution of regional endothermy and gigantism in the otodontid clade. *Historical Biology, 36(7)*, 1259–1273. https://doi.org/10.1080/08912963.2023.2211597

2.   McKenzie, R. W., Motta, P. J. & Rohr, J. R. (2014), Comparative squamation of the lateral line canal pores in sharks. *J Fish Biol, 84*, 1300-1311. https://doi.org/10.1111/jfb.12353

3.   Dillon, E. M., O'Dea, A. & Norris, R. D. (2017). Dermal denticles as a tool to reconstruct shark communities. *Marine Ecology Progress Series, 566*, pg. 117-134. https://doi.org/10.3354/meps12018

4.   Hobson, M. (2024). *Megalodon didn't look like a 50-foot great white shark, a controversial study claims.* Live Science. January 21, 2024. https://www.livescience.com/animals/extinct-species/controversial-study-claims-megalodon-didnt-look-like-a-50-foot-giant-great-white-shark

5.   Harvey, J. (2023). *What is a Boat Keel? & What Does It Do? – Detailed Explanation.* Boating Basics Online. October 1, 2023. https://www.boatingbasicsonline.com/boat-keel/

6.   Taylor & Francis. (2023). *Tiny scales reveal megalodon was not as fast as believed, but it had a mega-appetite explaining its gigantism.* PHYS.org. July 13, 2023. https://phys.org/news/2023-07-tiny-scales-reveal-megalodon-fast.html

7.   Cooper, J. A. et al. (2022). *The extinct shark Otodus megalodon was a transoceanic superpredator: Inferences from 3D modeling.* Science Advances. Published online August 17, 2022. https://doi.org/10.1126/sciadv.abm9424

8.   He, J., Tu, J., Yu, J. & Jiang, H. (2023). A global assessment of Bergmann's rule in mammals and birds. *Global change biology. 29(18).* https://doi.org/10.1111/gcb.16860

9.   Baker, H. (2022). *Megalodon was fastest swimming shark ever and could devour an orca in 5 bites, 3D model reveals.* Live Science. August 18, 2022. https://www.livescience.com/3d-megalodon-model

10.   Jacoby, D. M. P., Siriwat, P., Freeman, R. & Carbone, C. (2016). Correction to 'Is the scaling of swim speed in sharks driven by metabolism?' *Biol. Lett. 12*, 20160775. http://dx.doi.org/10.1098/rsbl.2016.0775

11.   Pyenson, N. D. & Koch, P. L. (2022). Oh, the shark has such teeth: Did megatooth sharks play a larger role in prehistoric food webs? *Science Advances. Vol. 8, Issue 25*, eadd2674. https://doi.org/10.1126/sciadv.add267

12. Watanabe, Y. Y., Goldman, K. J., Caselle, J. E., Chapman, D. D., & Papastamatiou, Y. P. (2015). Comparative analyses of animal-tracking data reveal ecological significance of endothermy in fishes. *Proceedings of the National Academy of Sciences, 112(19)*, 6104-6109. https://doi.org/10.1073/pnas.1500316112

13. Del Raye, G., Jorgensen, S. J., Krumhansl, K., Ezcurra, J. M., & Block, B. A. (2013). Travelling light: white sharks (Carcharodon carcharias) rely on body lipid stores to power ocean-basin scale migration. *Proceedings of the Royal Society B: Biological Sciences, 280(1766)*, 20130836. https://doi.org/10.1098/rspb.2013.0836

14. Zacharias, J. (2018). *Giant Megalodon Shark*. Museum of Arts & Sciences. October 18, 2018. https://www.moas.org/Giant-Megalodon-Shark-1-51.html

15. McDonald, A. (2023). *Scientists find new clue in what led to megalodon's demise*. CNN. July 11, 2023. https://edition.cnn.com/2023/07/03/world/megatooth-shark-warm-blood-scn/#:~:text=Like%20modern%20great%20white%20and%20mako%20sharks%2C%20megalodons,regulated%20by%20the%20temperature%20of%20water%20around%20them.

16. University of Zurich. (2022). *New 3D model shows: Megalodon could eat prey the size of entire killer whales*. PHYS.org. August 19, 2022. https://phys.org/news/2022-08-3d-megalodon-prey-size-entire.html#:~:text=It%20was%20estimated%20that%20it%20swam%20at%20around,eating%20whole%20prey%20up%20to%208%20meters%20long.

17. Goldbogen, J. A., Calambokidis, J., Shadwick, R. E., Oleson, E. M., McDonald, M. A. & Hildebrand, J. A. (2006). Kinematics of foraging dives and lunge-feeding in fin whales. *Journal of Experimental Biology, 209 (7)*, 1231–1244. https://doi.org/10.1242/jeb.02135

18. Sternes, P. C., et al. (2024). *White shark comparison reveals a slender body for the extinct megatooth shark, Otodus megalodon (Lamniformes: Otodontidae)*. Palaeontologia Electrnoica, Article number: 27.1.a7. https://doi.org/10.26879/1345

19. Bonfil, R., et al. (2005). Transoceanic Migration, Spatial Dynamics, and Population Linkages of White Sharks. *Science, 310*, 100-103 https://doi.org/10.1126/science.1114898

20. Shimada, K. et al. (2025). Reassessment of the possible size, form, weight, cruising speed, and growth parameters of the extinct megatooth shark, Otodus megalodon (Lamniformes: Otodontidae), and new evolutionary insights into its gigantism, life history strategies, ecology, and extinction. *Palaeontologia Electronica, 28(1)*:a12. https://doi.org/10.26879/1502

21. Fish, F. E. (2018). *Streamlining*. Encyclopedia of Marine Mammals (Third Edition), Academic Press. Pages 951-954. https://doi.org/10.1016/B978-0-12-804327-1.00250-8

22. Ahlborn, Boye & Blake, Robert & Chan, Keith. (2009). Optimal fineness ratio for minimum drag in large whales. *Canadian Journal of Zoology*. 87, 124-131. https://doi.org/10.1139/Z08-144

Chapter 5. Global Traveler References

1. Henkes, G. A. (2023). Dental geochemistry reveals thermoregulation in the Neogene ocean's most infamous super predator. *Proceedings of the National Academy of Sciences, Vol. 120, No. 27*, https://doi.org/10.1073/pnas.2308015120

2. Hayes, J.P. & Garland, T., Jr. (1995), The Evolution of Endothermy: Testing the Aerobic Capacity Model. *Evolution, 49*, 836-847. https://doi.org/10.1111/j.1558-5646.1995.tb02320.x

3. Tomita, T., Murakumo, K., & Matsumoto, R. (2023). Narrowing, twisting, and undulating: complicated movement in shark spiral intestine inferred using ultrasound. *Zoology, 157*:126077. https://doi.org/10.1016/j.zool.2023.126077

4. Dickson, K. & Graham, J. B. (2004). Evolution and consequences in endothermy in fishes. *Physiological Biochemical Zoology, 77*:998–1018. https://doi.org/10.1086/423743

5. Leigh, S. C., Summers, A. P., Hoffmann, S. L., & German, D. P. (2021). Shark spiral intestines may operate as Tesla valves. *Proceedings of the Royal Society B, 288*:20211359. https://doi.org/10.1098/rspb.2021.1359

6. Bernal, D., Dickson, K. A., Shadwick, R. E., & Graham, J. B. (2001). Review: analysis of the evolutionary convergence for high-performance swimming in lamnid sharks and tunas. *Comparative Biochemistry and Physiology Part A: Molecular and Integrative Physiology, 129*:695–726. https://doi.org/10.1016/S1095-6433(01)00333-6

7. Razak, H. & Kocsis, L. (2018). Late Miocene Otodus (Megaselachus) megalodon from Brunei Darussalam: body length estimation and habitat reconstruction. *Neues Jahrbuch für Geologie und Paläontologie – Abhandlungen, 288*:299–306. https://doi.org/10.1127/njgpa/2018/0743

8. French, G.C.A., Stürup, M., Rizzuto, S., van Wyk, J.H., Edwards, D., Dolan, R.W., Wintner, S.P., Towner, A.V. & Hughes, W.O.H. (2017). The tooth, the whole tooth, and nothing but the tooth: tooth shape and ontogenetic shift dynamics in the white shark Carcharodon carcharias. *J Fish Biol, 91*, 1032-1047. https://doi.org/10.1111/jfb.13396

9. Griffiths, M., Eagle, R., Kim, S., Flores, R., Becker, M., Maisch IV, H., Trayler, R., Chan, R., McCormack, J., Akhtar, A., Tripati, A. & Shimada, K. (2023). Endothermic physiology of extinct megatooth sharks. *PNAS, Vol 120, No 27*, e2218153120. https://doi.org/10.1073/pnas.2218153120

10. McDonald, A. (2023). *Scientists find new clue in what led to megalodon's demise.* CNN. July 11, 2023. https://edition.cnn.com/2023/07/03/world/megatooth-shark-warm-blood-scn/#:~:text=Like%20modern%20great%20white%20and%20mako%20sharks%2C%20megalodons,regulated%20by%20the%20temperature%20of%20water%20around%20them.

11. Carlson, J. K., Goldman, K. J., & Lowe, C. G. (2004). Metabolism, energetic demand, and endothermy. *Biology of sharks and their relatives, 10*, 269-286. https://doi.org/10.1201/9780203491317.ch7

12. Pyenson, N. D. & Koch, P. L. (2022). Oh, the shark has such teeth: Did megatooth sharks play a larger role in prehistoric food webs? *Science Advances. Vol. 8, Issue 25*, eadd2674. https://doi.org/10.1126/sciadv.add267

13. Cooper, J. A. et al. (2022). *The extinct shark Otodus megalodon was a transoceanic superpredator: Inferences from 3D modeling.* Science Advances. Published online August 17, 2022. https://doi.org/10.1126/sciadv.abm9424

14.   Bendix-Almgreen, S. E. (1983). Carcharodon megalodon from the Upper Miocene of Denmark, with comments on elasmobranch tooth enameloid: coronoin. *Bulletin of the Geological Society of Denmark, Vol 32, No. 1-2*, pp. 1-32. https://doi.org/10.37570/bgsd-1983-32-01

15.   Shimada, K., Maisch, H. M., Perez, V. J., Becker, M. A., & Griffiths, M. L. (2022). Revisiting body size trends and nursery areas of the Neogene megatooth shark, Otodus megalodon (Lamniformes: Otodontidae), reveals Bergmann's rule possibly enhanced its gigantism in cooler waters. *Historical Biology, 35*(2), 208–217. https://doi.org/10.1080/08912963.2022.2032024

16.   Sottile, Z. (2022). *The extinct superpredator megalodon was big enough to eat orcas, scientists say.* CNN Digital. August 20, 2022. https://www.ctvnews.ca/sci-tech/the-extinct-superpredator-megalodon-was-big-enough-to-eat-orcas-scientists-say-1.6035361#:~:text=An%20adult%20megalodon%20would%20have%20needed%20to%20eat,times%20higher%20than%20an%20adult%20great%20white%20shark.

17.   He, J., Tu, J., Yu, J. & Jiang, H. (2023). A global assessment of Bergmann's rule in mammals and birds. *Global change biology. 29(18)*. https://doi.org/10.1111/gcb.16860

18.   Meiri, S. & Dayan, T. (2003). On the validity of Bergmann's Rule. *Journal of Biogeography, 30(3)*, 331–351. https://doi.org/10.1046/j.1365-2699.2003.00837.x

19.   Pimiento, C., MacFadden, B.J., Clements, C.F., Varela, S., Jaramillo, C., Velez-Juarbe, J. & Silliman, B.R. (2016), Geographical distribution patterns of Carcharocles megalodon over time reveal clues about extinction mechanisms. J. Biogeogr., 43, 1645-1655. https://doi.org/10.1111/jbi.12754

20.   Ferrón, H. G. (2017). Regional endothermy as a trigger for gigantism in some extinct macropredatory sharks. PLoS ONE, 12(9), e0185185. https://doi.org/10.1371/journal.pone.0185185

21.   Pimiento, C., Cantalapiedra, J. L., Shimada, K., Field, D. J., & Smaers, J. B. (2019). Evolutionary pathways toward gigantism in sharks and rays. *Evolution, 73 (2)*, 588–599. https://doi.org/10.1111/evo.13680

22.   Dickson, K. A., & Graham, J. B. (2004). Evolution and Consequences of Endothermy in Fishes. Physiological and Biochemical Zoology: *Ecological and Evolutionary Approaches, 77(6)*, 998–1018. https://doi.org/10.1086/423743

Chapter 6. Fossil Evidence of Attacks References

1.   Anderson J.M. et al. (2021). Non-random co-occurrence of juvenile white sharks (Carcharodon carcharias) at seasonal aggregation sites in southern California. *Front. Mar. Sci. 8*, 688505. https://doi.org/10.3389/fmars.2021.688505

2.   Schilds, A., Mourier, J., Huveneers, C., Nazimi, L., Fox, A. & Leu, S. T. (2019). Evidence for non-random co-occurrences in a white shark aggregation. *Behav. Ecol. Sociobiol. 73*, 138. https://doi.org/10.1007/s00265-019-2745-1

3.   Papastamatiou, Y. P., Mourier, J., TinHan, T., Luongo, S., Hosoki, S., Santana-Morales, O., & Hoyos-Padilla, M. (2022). Social dynamics and individual hunting tactics of white sharks revealed by biologging. *Biology Letters, 18*(3). https://doi.org/10.1098/rsbl.2021.0599

4.   Klimley, A. P., Le Boeuf, B. J., Cantara, K. M., Richert, J. E., Davis, S. F., Van Sommeran, S., Kelly, J.T. (2001). The hunting strategy of white sharks (Carcharodon carcharias) near a seal colony. *Mar. Biol. 138*, 617-636. https://doi.org/10.1007/s002270000489

5.   Papastamatiou, Y. P., Bodey, T. W., Caselle, J. E., Bradley, D., Freeman, R., Friedlander, A. M. & Jacoby, D. M. P. (2020). Multiyear social stability and social information use in reef sharks with diel fission-fusion dynamics. *Proc. R. Soc. B, 287*, 20201063. http://doi.org/10.1098/rspb.2020.1063

6.   Hoyos-Padilla, E. M., Klimley, A. P., Galván- Magaña, F. & Antoniou, A. (2016). Contrasts in the movements and habitat use of juvenile and adult white sharks (*Carcharodon carcharias*) at Guadalupe Island, Mexico. *Anim. Biotelem. 4*, 14. https://doi.org/10.1186/s40317-016-0106-7

7.   Skomal, G. B., Hoyos-Padilla, E. M., Kukulya, A. & Stokey, R. (2015). Subsurface observations of white shark *Carcharodon carcharias* predatory behavior using an autonomous underwater vehicle. *J. Fish Biol. 87*, 1293-1312. https://doi.org/10.1111/jfb.12828

8.   Towner, A. V., Leos-Barajas, V., Langrock, R., Schick, R. S., Smale, M. J., Kaschke, T., Jewell, O. J. D. & Papastamatiou, Y. P. (2016). Sex-specific and individual preferences for hunting strategies in white sharks. *Funct. Ecol. 30*, 1397-1407. https://doi.org/10.1111/1365-2435.12613

9.   Watanabe, Y. Y., Payne, N. L., Semmens, J. M., Fox, A. & Huveneers, C. (2019). Hunting behavior of white sharks recorded by animal-borne accelerometer and cameras. *Mar. Ecol. Prog. Ser. 621*, 221-227. https://doi.org/10.3354/meps12981

10.   Aplin, L. M., Farine. D. R., Morand-Ferron, J., Cole, E. F., Cockburn, A. & Sheldon, B.C. (2013). Individual personalities predict social behavior in wild networks of great tits (Parus major). *Ecol. Lett. 16*, 1365-1372. https://doi.org/10.1111/ele.12181

11.   Godfrey, S. J. & Beatty, B. L. (2022). A Shear-Compression Fracture. *Society of Vertebrate Paleontology. Article number: 25.3.a28*. https://doi.org/10.26879/1171

12.   Krisch, J. A. (2022). *Terrifying megalodon attack on whale revealed in 15-million-year-old fossils*. Live Science. September 9, 2022. https://www.livescience.com/megalodon-vs-whale-failed-predation

13.   Godfrey, S. J., Ellwood, M., Groff, S. & Verdin, M. S. (2018). Carcharocles-bitten odontocete caudal vertebrae from the Coastal Eastern United States. *Acta Palaeontologica Polonica, 63 (3)*, 463-468. https://doi.org/10.4202/app.00495.2018

14.   Collareta, A., Lambert, O., Landini, W., et al. (2017). Did the giant extinct shark Carcharocles megalodon target small prey? Bite marks on marine mammal remains from the late Miocene of Peru. *Palaeogeography, Palaeoclimatology, Palaeoecology, Volume 469*, Pages 84-91. https://doi.org/10.1016/j.palaeo.2017.01.001

15. Cooper, J. A. et al. (2022). *The extinct shark Otodus megalodon was a transoceanic superpredator: Inferences from 3D modeling.* Science Advances. Published online August 17, 2022. https://doi.org/10.1126/sciadv.abm9424

16. Bianucci, G., Sorce, B., Storai, T. & Landini, W. (2010). Killing in the Pliocene: Shark attack on a dolphin from Italy. *Palaeontology, 53(2)*, 457-470. https://doi.org/10.1111/j.1475-4983.2010.00945.x

17. Kallal, R.J., Godfrey, S.J. and Ortner, D.J. (2012), Bone reactions on a Pliocene cetacean rib indicate short-term survival of predation event. *Int. J. Osteoarchaeol., 22(3)*, 253-260. https://sci-hub.st/10.1002/oa.1199

18. Mierzwiak, J. S. & Godfrey, S. J. (2019). Megalodon-bitten Whale Rib From South Carolina. *Ecphora, Volume 34, Number 2*, Pages 15 – 20. https://www.calvertmarinemuseum.com/DocumentCenter/View/3408/Ecphora-June-2019

19. Godfrey, Stephen & Nance, John & Riker, Norman. (2021). Carcharocles-bitten sperm whale tooth from the Neogene of the Coastal Eastern United States. *Acta Palaeontologica Polonica, 66(3)*, https://doi.org/10.4202/app.00820.2020

20. Godfrey, S., Bennett, M., & Perez, V. (2024). Taphonomic and ecological insights from conspecific bite marks on Otodus megalodon teeth. *Acta Palaeontologica Polonica, 69 (4)*, 731-736. https://doi.org/10.4202/app.01188.2024

21. Borucinska, J., Adams, D. & Frazier, B. (2020). Histologic observations of dermal wound healing in a free-ranging blacktip shark Carcharhinus limbatus from the southeastern U.S. Atlantic coast: a case report. *Journal of Aquatic Animal Health, 32(4)*. https://doi.org/10.1002/aah.10113

22. Chin, A., Mourier, J., & Rummer, J. L. (2015). Blacktip reef sharks (Carcharhinus melanopterus) show high capacity for wound healing and recovery following injury. *Conservation physiology, 3(1)*, cov062. https://doi.org/10.1093/conphys/cov062

23. Womersley, F., Hancock, J., Cameron, T. P. & Rowat, D. (2021). Wound-healing capabilities of whale sharks (Rhincodon typus) and implications for conservation management. *Conservation Physiology, Volume 9, Issue 1, coaa120*, 16 pages. https://doi.org/10.1093/conphys/coaa120

24. Black, C. (2023). Resilience in the Depths: First Example of Fin Regeneration in a Silky Shark (Carcharhinus falciformis) following Traumatic Injury. *Journal of Marine Sciences, Vol 2023, Issue 1, Article 6639805*, 8 pages. https://doi.org/10.1155/2023/6639805

25. Woodward, A. & Business Insider. (2019). *Scientists Just Mapped The Great White Shark's Genome in Search For Healing Powers.* Nature. February 21, 2019. https://www.sciencealert.com/scientists-just-mapped-the-great-white-shark-s-genome-in-search-for-healing-powers

26. Huveneers, C., Klebe, S., Fox, A., Bruce, B., Robbins, R., Borucinska, J., Jones, R. & Michael, M. (2016). First histological examination of a neoplastic lesion from a free-swimming white shark, Carcharodon carcharias L. *Journal of fish diseases, 39(10)*, https://doi.org/10.1111/jfd.12458

27. Marra, N. J. el al. (2019). White shark genome reveals ancient elasmobranch adaptations associated with wound healing and the maintenance of genome stability. *PNAS, 116(10)*, 4446-4455. https://doi.org/10.1073/pnas.1819778116

28. Luer C. A. & Walsh, C. J. (2018). Potential Human Health Applications from Marine Biomedical Research with Elasmobranch Fishes. *Fishes, 3(4)*, 47. https://doi.org/10.3390/fishes3040047

Chapter 7. Megalodon Teeth References

1. Alden, Andrew. (2021, February 16). *Official State Dinosaurs and Fossils*. Retrieved from https://www.thoughtco.com/official-state-fossils-and-dinosaurs-1441148

2. Coates, L. (2022). *Sammy, 6, finds 'once-in-a-lifetime' rare fossil on beach*. Great Yarmouth Mercury. May 4, 2022. https://www.greatyarmouthmercury.co.uk/news/20991867.sammy-6-finds-once-in-a-lifetime-rare-fossil-beach/

3. Maisch, IV, H.M., Becker, M. A. & Chamberlain, J. A. (2018). Lamniform and Carcharhiniform Sharks from the Pungo River and Yorktown Formations (Miocene–Pliocene) of the Submerged Continental Shelf, Onslow Bay, North Carolina, USA. *Copeia, 106(2)*, 353-374. https://doi.org/10.1643/OT-18-016

4. https://www.theworldslargestsharksjaw.com/

5. https://www.fossilera.com/pages/what-is-the-largest-megalodon-tooth-ever-found

6. Cooper, J.A., Pimiento, C., Ferrón, H.G. et al. (2020). Body dimensions of the extinct giant shark Otodus megalodon: a 2D reconstruction. *Sci Rep, 10*, 14596 https://doi.org/10.1038/s41598-020-71387-y

7. French, G.C.A., Stürup, M., Rizzuto, S., van Wyk, J.H., Edwards, D., Dolan, R.W., Wintner, S.P., Towner, A.V. & Hughes, W.O.H. (2017). The tooth, the whole tooth, and nothing but the tooth: tooth shape and ontogenetic shift dynamics in the white shark Carcharodon carcharias. *J Fish Biol, 91*, 1032-1047. https://doi.org/10.1111/jfb.13396

8. Perez, V. L., Leder, R. M. & Badaut, T. (2021). Body length estimation of Neogene macrophagous lamniform sharks (Carcharodon and Otodus) derived from associated fossil dentitions. *Palaeologica Electronica. Article number: 24.1.a09*. https://doi.org/10.26879/1140

9. Boessenecker, R. W., Ehret, D. J., Long, D. J., Churchill, M., Martin, E, & Boessenecker, S. J. (2019). The Early Pliocene extinction of the mega-toothed shark Otodus megalodon: a view from the eastern North Pacific. *Peer J, 7*:e6088. https://doi.org/10.7717/peerj.6088

10. Shimada, K. (2019). The size of the megatooth shark, Otodus megalodon (Lamniformes: Otodontidae), revisited. *Historical Biology, 33*, 904 - 911. https://doi.org/10.1080/08912963.2019.1666840

11. Shimada, K. et al. (2025). Reassessment of the possible size, form, weight, cruising speed, and growth parameters of the extinct megatooth shark, *Otodus megalodon* (Lamniformes: Otodontidae), and new evolutionary insights into its gigantism, life history strategies, ecology, and extinction. *Palaeontologia Electronica, 28(1)*:a12. https://doi.org/10.26879/1502

12. Tucker, A. S. & Fraser, G. J. (2014). Evolution and developmental diversity of tooth regeneration. *Seminars in cell & developmental biology, 25-26*, 71-80. https://doi.org/10.1016/j.semcdb.2013.12.013

13. Rasch, Liam & Martin, Kyle & Cooper, Rory & Metscher, Brian & Underwood, Charlie & Fraser, Gareth. (2016). An ancient dental gene set governs development and continuous regeneration of teeth in sharks. *Developmental Biology, 415(2)*. 10.1016/j.ydbio.2016.01.038

14. https://facts.net/megalodon-facts/

15. Skinner, A. (2022). *A Look Back at Megalodon Discoveries From The Last 115 Years*. Newsweek. September 7, 2022. https://www.newsweek.com/look-back-megaladon-discoveries-115-years-1740721

16. Bowling, T. & Fraser, G. (2020). *A Shark's Infinite Regeneration of Teeth: An Interview with UF Biology's own Dr. Gareth Fraser*. Florida Museum. February 17, 2020. https://www.floridamuseum.ufl.edu/sharks/blog/a-sharks-infinite-regeneration-of-teeth/#:~:text=Sharks%20do%20not%20rely%20on%20two%20sets%20of,new%20set%20of%20teeth%20develops%20every%20two%20weeks%21

17. Davis, J. (2025). *Megalodon: The truth about the largest shark that ever lived*. Natural History Museum. https://www.nhm.ac.uk/discover/megalodon--the-truth-about-the-largest-shark-that-ever-lived.html

18. Heithaus, M. (2022). *Millions of years ago, the megalodon ruled the oceans – why did it disappear?* Florida International University. June 20, 2022. https://news.fiu.edu/2022/millions-of-years-ago,-the-megalodon-ruled-the-oceans-why-did-itdisappear#:~:text=Megalodons%20are%20extinct.,they%20looked%20at%20the%20teeth

Chapter 8. Variations in Megalodon Teeth References

1. Maisch, IV, H.M., Becker, M. A. & Chamberlain, J. A. (2018). Lamniform and Carcharhiniform Sharks from the Pungo River and Yorktown Formations (Miocene–Pliocene) of the Submerged Continental Shelf, Onslow Bay, North Carolina, USA. *Copeia, 106(2)*, 353-374. https://doi.org/10.1643/OT-18-016

2. Becker, M. A., Chamberlain Jr., J. A. & Stoffer, P. W. (2000). Pathologic tooth deformities in modern and fossil chondrichthians: a consequence of feeding related injury. *Lethaia, Vol 33, Issue 2*, pp 103-118. https://doi.org/10.1080/00241160050150249

3. Miller, H. S., Avrahami, H. M., & Zanno, L. E. (2022). Dental pathologies in lamniform and carcharhiniform sharks with comments on the classification and homology of double tooth pathologies in vertebrates. *PeerJ, 10*, e12775. https://doi.org/10.7717/peerj.12775

4. Hunasgi, S., Koneru, A., Manvikar, V., Vanishree, M. & Amrutha, R. (2017). A Rare Case of Twinning Involving Primary Maxillary Lateral Incisor with Review of Literature. *J Clin of Diagn Res.* 11(2), ZD09-ZD11. https://www.doi.org/10.7860/JCDR/2017/23510/9212

5. Clark, E. J., Chesnutt, S. R., Winer, J. N., Kass, P. H., & Verstraete, F. J. M. (2017). Dental and Temporomandibular Joint Pathology of the American Black Bear (Ursus americanus). *Journal of Comparative Pathology, 156(2-3)*, 240–250. https://doi.org/10.1016/j.jcpa.2016.11.267

6. Kahle, P., Ludolphy, C., Kierdorf, H., & Kierdorf, U. (2018). Dental anomalies and lesions in Eastern Atlantic harbor seals, Phoca vitulina vitulina (Carnivora, Phocidae), from the German North Sea. *PLOS ONE, 13(10)*, e0204079. https://doi.org/10.1371/journal.pone.0204079

7. Loch. C., Grando, L. J., Kieser,J. A. & Simões-Lopes, P. C. (2011) Dental pathology in dolphins (Cetacea: Delphinidae) from the southern coast of Brazil. *Dis Aquat Org, 94(3)*, 225-234. https://doi.org/10.3354/dao02339

8. Burns, J., Baker, C. G. & Mol, D. (2003). An extraordinary woolly mammoth molar from Alberta, Canada. *Deinsea, 9(1)*, 77–86. https://natuurtijdschriften.nl/pub/538672

9. Kamura, Y. (2019). *Variations in the Anatomy of the Teeth. In:* Iwanaga, J., Tubbs, R. (eds) Anatomical Variations in Clinical Dentistry. Springer, Cham. https://doi.org/10.1007/978-3-319-97961-8_20

10. Cetinbas, T., Halil, S., Akcam, M. O., Sari, S., & Cetiner, S. (2007). Hemisection of a fused tooth. *Oral surgery, oral medicine, oral pathology, oral radiology, and endodontics, 104(4)*, e120–e124. https://doi.org/10.1016/j.tripleo.2007.03.029

11. Kryukova, N. (2017). Cases of teeth concrescence in the Pacific walrus (Odobenus rosmarus divergens). *Russian Journal of Theriology, 16(1)*, 110-113. https://doi.org/10.15298/rusjtheriol.16.1.10

12. Tsesis, I., Steinbock, N., Rosenberg, E. & Kaufman, A.Y. (2003). Endodontic treatment of developmental anomalies in posterior teeth: treatment of geminated/fused teeth – report of two cases. *International Endodontic Journal, 36(5)*, 372-379. https://doi.org/10.1046/j.1365-2591.2003.00666.x

Chapter 9. Why Did Megalodon Go Extinct? References

1. Pimiento, C., MacFadden, B.J., Clements, C.F., Varela, S., Jaramillo, C., Velez-Juarbe, J. & Silliman, B.R. (2016), Geographical distribution patterns of Carcharocles megalodon over time reveal clues about extinction mechanisms. *J. Biogeogr., 43*, 1645-1655. https://doi.org/10.1111/jbi.12754

2. Pimiento, C., Griffin, J. N., Clements, C. F., Silvestro, D., Varela, S., Uhen, M. D., & Jaramillo, C. (2017). The Pliocene marine megafauna extinction and its impact on functional diversity. *Nature Ecology & Evolution, 1(8)*, 1100-1106. https://doi.org/10.1038/s41559-017-0223-6

3. Herbert, T., Lawrence, K., Tzanova, A., et al. (2016). Late Miocene global cooling and the rise of modern ecosystems. *Nature Geoscience, 9*, 843–847. https://doi.org/10.1038/ngeo2813

4.   Brown, R.M., Chalk, T.B., Crocker, A.J., et al. (2022). Late Miocene cooling coupled to carbon dioxide with Pleistocene-like climate sensitivity. *Nat. Geosci.* 15, 664–670. https://doi.org/10.1038/s41561-022-00982-7

5.   Klaus, J. S., et al. (2011). Rise and Fall of Pliocene Free-Living Corals in the Caribbean. *Geology, 39(4)*, 375–378. https://doi.org/10.1130/G31704.1

6.   Tanner, T., Hernández-Almeida, I., Drury, A.J., Guitián, J. & Stoll, H. (2020). Decreasing Atmospheric CO2 During the Late Miocene Cooling. *Paleoceanography and Paleoclimatology, Volume 35, Issue 12*, e2020PA003925, 1-25. https://doi.org/10.1029/2020PA003925

7.   Harnik, P., et al. (2012). Extinctions in ancient and modern seas. *Trends in Ecology & Evolution, 27, 608-617.* https://doi.org/10.1016/j.tree.2012.07.010

8.   Sinha, D.K., Singh, A.K. (2021). Foraminiferal micropaleontology for understanding Earth's history. *J Earth Syst Sci, Volume 130, Article Number 225*, Pages 281-319. https://doi.org/10.1007/s12040-021-01735-7

9.   Guillermic, M., Misra, S., Eagle, R., and Tripati, A.: Atmospheric CO2 estimates for the Miocene to Pleistocene based on foraminiferal $\delta^{11}B$ at Ocean Drilling Program Sites 806 and 807 in the Western Equatorial Pacific, *Clim. Past, Volume 18, Issue 2*, 183–207, https://doi.org/10.5194/cp-18-183-2022

10. Lear, C. H., Elderfield, H. & Wilson, P.A. (2000). Cenozoic Deep-Sea Temperatures and Global Ice Volumes from Mg/Ca in Benthic Foraminiferal Calcite. *Science, 287*, 269-272. https://doi.org/10.1126/science.287.5451.269

11. Fehrenbacher, J., Russell, A., Davis, C. et al. (2017). Link between light-triggered Mg-banding and chamber formation in the planktic foraminifera Neogloboquadrina dutertrei. *Nat Commun, Volume 8, Article 15441.* https://doi.org/10.1038/ncomms15441

12. Gooday, A.J. (2001). *Benthic Foraminifera.* Editor(s): John H. Steele, Encyclopedia of Ocean Sciences, Academic Press, Pages 274-286, https://doi.org/10.1006/rwos.2001.0217.

13. Estes, J. A., et al. (2011). Trophic Downgrading of Planet, *Earth Science, 333(6040)*, 301-306. https://doi.org/10.1126/science.1205106

14. Martin, J. E., Tacail, T., Adnet, S., Girard, C. & Balter, V. (2015). Calcium isotopes reveal the trophic position of extant and fossil elasmobranchs. *Chemical Geology, Volume 415*, Pages 118-125. https://doi.org/10.1016/j.chemgeo.2015.09.011

15. Dickson, K. A., & Graham, J. B. (2004). Evolution and Consequences of Endothermy in Fishes. Physiological and Biochemical Zoology: *Ecological and Evolutionary Approaches, 77(6)*, 998–1018. https://doi.org/10.1086/423743

16. McCormack, J., Griffiths, M. L., Kim, S. L., Kenshu, S., Karnes, M., Maisch IV, H. et al. (2022). Trophic position of Otodus megalodon and great white sharks through time revealed by zinc isotopes. *Nat Commun 13*, 2980. https://doi.org/10.1038/s41467-022-30528-9

17. Kast, E., Griffiths, M., & Kim, S. & Rao, Z., et al. & (2022). Cenozoic megatooth sharks occupied extremely high trophic positions. *Science Advances, 8(25)*, eabl6529. https://doi.org/10.1126/sciadv.abl6529

18. Perez, V. J., Godfrey, S. J., Kent, B. W., Weems, R. E., & Nance, J. R. (2018). The transition between Carcharocles chubutensis and Carcharocles megalodon (Otodontidae, Chondrichthyes): lateral cusplet loss through time. *Journal of Vertebrate Paleontology, 38(6)*. https://doi.org/10.1080/02724634.2018.1546732

19. Nyberg, K. G., Ciampaglio, C. N., & Wray, G. A. (2006). Tracing the ancestry of the great white shark, Carcharodon carcharias, using morphometric analyses of fossil teeth. *Journal of Vertebrate Paleontology, 26(4)*, 806-814. https://doi.org/10.1671/0272-4634(2006)26[806:ttaotg]2.0.co;2

20. Haiken, M. (2024). *Baby great white shark reveals huge nursery near NYC in scientific first.* National Geographic. June 24, 2024. https://www.nationalgeographic.com/animals/article/babies-great-white-sharks-camera-new-york

21. Ashworth, J. (2022). *Megalodon sharks grew biggest in colder waters.* Natural History Museum. March 7, 2022. https://www.nhm.ac.uk/discover/news/2022/march/megalodon-sharks-grew-biggest-colder-waters.html

22. Griffiths, M., Eagle, R., Kim, S., Flores, R., Becker, M., Maisch IV, H., Trayler, R., Chan, R., McCormack, J., Akhtar, A., Tripati, A. & Shimada, K. (2023). Endothermic physiology of extinct megatooth sharks. *PNAS, Vol 120, No 27*, e2218153120. https://doi.org/10.1073/pnas.2218153120

23. Boessenecker, R. W., Ehret, D. J., Long, D. J., Churchill, M., Martin, E, & Boessenecker, S. J. (2019). The Early Pliocene extinction of the mega-toothed shark Otodus megalodon: a view from the eastern North Pacific. *Peer J,* 7:e6088. https://doi.org/10.7717/peerj.6088

Chapter 9. Does Megalodon Still Exist? References

1. Joseph, J. (2025). *Research ship finds sea creature, untouched since dinosaur times, after years of searching.* Eath.com. January 20, 2025. https://www.earth.com/news/research-ship-finds-palau-nautilus-sea-creature-untouched-since-dinosaur-times/

2. Barord, G., Combosch, D., Giribet, G., Landman, N., Lemer, S., Veloso, Job L., & Ward, P. (2023). *Three new species of Nautilus Linnaeus, 1758 (Mollusca, Cephalopoda) from the Coral Sea and South Pacific.* ZooKeys. 1143(3). 51-69. https://doi.org/10.3897/zookeys.1143.84427

3. Urton, J. (2015). *Rare nautilus sighted for the first time in three decades.* U.W. News. August 25, 2015. https://www.washington.edu/news/2015/08/25/rare-nautilus-sighted-for-the-first-time-in-three-decades/

4. Lewis, D. (2015). *Marine Biologists Find Rare Nautilus For The First Time In 30 Years.* Smithsonian Magazine. August 31, 2015. https://www.smithsonianmag.com/smart-news/marine-biologist-stumbles-rare-nautilus-first-time-30-years-180956445/

5. Priede, I. G., et al. (2006). The absence of sharks from abyssal regions of the world's oceans. *Proc. R. Soc. B.273*, 1435–1441. http://doi.org/10.1098/rspb.2005.3461

6. Pollerspöck, J., Cares, D., Ebert, D. A., Kelley, K. A., Pockalny, R., Robinson, R. S., ... Straube, N. (2023). First in situ documentation of a fossil tooth of the megatooth shark Otodus (Megaselachus) megalodon from the deep sea in the Pacific Ocean. *Historical Biology*, 1–6. https://doi.org/10.1080/08912963.2023.2291771

7. Lambert, O., Bianucci, G., Post, K. *et al.* (2010). The giant bite of a new raptorial sperm whale from the Miocene epoch of Peru. *Nature, 466*, 105–108. https://doi.org/10.1038/nature09067

8. Greenfield, T. (2023). Of megalodons and men: Reassessing the 'modern survival' of Otodus megalodon. *Journal of Scientific Exploration. 37*, 330-347. https://doi.org/10.31275/20233041

Surviving Undetected in the Open Seas. © iStock. Credit: Alessandro De Maddalena.

Bone Valley *Otodus megalodon* on Kaleidoscope Jasper.
Courtesy of Dr. Adam Bozeman.

INDEX

C

D

Is Megalodon still out there?
Courtesy of Alessandro De Maddalena.